Animals in the Service of Man

10,000 YEARS OF DOMESTICATION

Edward Hyams

J. M. Dent & Sons Ltd, London

Contents

Illustrations

Plates 1–11, 13, 14, 16, 19, 20, 29, 30, 34–7, 39, 41, 42, the Zoological Society of London;

Pls. 7 and 28 by courtesy of the French Government Tourist Office;

Pls. 12, 15 (photo, Peter Jackson), 17, 18, 21 (photos, Jane Burton), 22 (photo, S. G. Porter), 32, 33, Bruce Coleman Ltd;

Pls. 23, 38, 40, Popperfoto Library;

Pls. 24, 25 by courtesy of Danish Agricultural Producers;

Pls. 26, 27, Copyright by Berlin Zoo;

Pl. 31a, b, c, The Natural History Photographic Agency; photographs by Stephen Dalton, F.R.P.S.

Author's Preface

The beginning of civilization is plant and animal domestication.

Some Palaeolithic peoples, living by food-gathering and hunting, showed technological and artistic genius, for example the Cro-Magnons of the Magdalenian. But for all that they could never, by such means of getting their living, have accumulated the surplus required to free some men and women from the labour of food-getting and so leave them at liberty to turn their minds and hands to other work; to the devising of new techniques, the building of cities, the development of philosophies, sciences and arts and the refinement of religions. Domestication of plants and animals emancipated a proportion of the human race by enabling them to live on food grown by their fellows, purchased with labour in other kinds of work. And the proportion of our species engaged in food production has steadily declined because the contribution of the others to the common wealth has included the advancement of food-producing technology as well as the enrichment of life by industries and arts.

But our remote ancestors did not, with such results in mind, conceive the idea of domesticating plants and animals. There was nothing deliberate about it, it was not initiated by a stroke of genius, a brainwave. Domestication should be seen as evolution, not an event but a process; a process composed of an infinite number of small modifications in man's behaviour towards certain animals, and in theirs to him. The occurrence, over a long period of time, and by degrees, of the domestication of certain animals, itself pushed

man forward, opened up new perspectives, led man to modify
his own behaviour in order to meet a future he unwittingly
made possible when he slowly changed those animals from
free prey into bond slaves.

There is not, unfortunately, an historical method analogous
to the differential calculus in mathematics. We can study
man's relations with the rest of the universe only in representations of very coarse texture because we have no means
of exposing a continuum of point-events and are obliged to
represent it in discontinuous measurable events which are
not in nature.

But we do have an artistic method of suggesting the truth
we apprehend but cannot describe: impressionism. I will
take, from the chapters which follow, the case of the dog.
The wolf and the jackal are scavengers as well as hunters;
they make their way into man's camps and clean up what he
has left over from the kill of game (wild jackals still do so even
in cities). A mutually useful social relationship is started. A
boy and a wolf-cub play together and love each other; the
cub is reared tame; it finds a mate in the wild and persuades
her to join him; a breed of tame wolves is begun; in-breeding
gives expression to genetically recessive attributes in morphological changes, and so the dog comes into being. The
man-pack, keeping track of a moving herd of game animals
as it makes its way from pasture to pasture, cutting out and
killing stragglers, or driving a group of beasts into a contrived
or natural cul-de-sac trap, finds a wolf-pack doing much the
same; an accident, some move of the wolf-pack which so turns
the animals that the men get the chance they have been watch-
ing for, suggests mutual advantages, after the kill has been
shared, in joint hunting. Men and wolves, at first inadver-
tently, then deliberately, drive and corral game beasts. It is
the germ of the shepherd-sheepdog symbiosis. So, three, out
of a possible many, origins of the domestic dog: mutual social-
economic service; love of young creatures as pets; similarity
of hunting techniques.

Another example: a small community of people living by hunting and gathering begins to follow a herd of ungulates on the move, killing a few individuals at need. It becomes a habit, a way of life, the men living as parasites on the herd. This parasitism enables them to thrive; their numbers increase, they have more leisure, by slow degrees they begin to assert themselves and, from merely following and taking toll of the herd, begin to guide and protect it; at long last to control it. In this evolution from parasitism to a measure of mastery they are helped by their dogs; and by the discovery arising out of observation of means by which the animals of the herd can be bribed into tame obedience, for example the offer and regular provision of salt. Thus pastoral nomadism evolves. Young animals are made pets, are tamed, and grow into adults which can be milked, bled as the Masai bleed their cattle and drink the blood, and even, in some cases, harnessed and ridden. It is important that this evolution of a symbiotic out of a parasitic relationship is perfectly possible before the evolution of plant-farming. It is believed, now, that the domestication of reindeer, goats and sheep was pre-agricultural.

Other domestications evolved very differently. In another book [1] I have tried to describe how men came to domesticate plants and so became farmers. The farmers were not alone in seeing the advantage of having cultivated crops; deer, cattle, buffalo, even elephants, were attracted by easy feeding and became crop-robbers. But in doing so they themselves suggested to man the means of their own enslavement. Thus the domestication of a number of other species of animals which could be useful to man is a by-product of the domestication of plants.

Finally, and relatively late, there are the deliberate domestications, the deliberate and thought-out taking under human control of certain animals, with a clear purpose in mind. That purpose was usually, but not invariably, an economic one.

[1] Hyams (1971).

The domestication of the Paradise Fish, of the Canary, of the Budgerigar, were all for pleasure; the domestication of snails and of dormice was to satisfy gourmets or gluttons rather than hungry people; the domestication of minks and sables was dictated by fashion rather than by real need.

The work of bringing animals into our service, which began perhaps 15,000 years ago, is not finished. One immense task confronting us—which, in view of the food/population ratio is now a necessity—is the farming and ultimate domestication of marine fishes. A start has been made on this task, and I have devoted most of my final chapter to describing it.

E. H.

1

Men and Dogs

Before about 10,000 B.C. man got his living by gathering vegetable foods such as roots, nuts, berries and leaves, as well as animal foods such as molluscs and grubs, and by hunting wild animals for the meat and raw materials of primitive industry which their bodies yielded, such as bone, horn and hide, and, of course, fur. I have described elsewhere [1] how, in course of time, the springing up of seedlings from seeds carelessly dropped in and near his camps taught man the idea of agriculture, the deliberate planting and protection of food plants, instead of reliance on finding them in the woods and on the prairies. In much the same way did the hunting of wild animals at last suggest the idea of taming them, so as to keep a store of meat and other valuable materials 'on the hoof' until it was wanted; and, at a later stage, to make use of animal muscle-power for transport and industry.

The first animal to be domesticated by man was not required as meat, but to help hunt for the meat provision. The very ancient association between men and dogs has its origin in the discovery that they had more interests in common than sources of antipathy, and in the fact that they also had in common a certain kind of social organization—hunting in packs under an acknowledged leader.

The ancestry of *Canis familiaris*, the species to which all the many races of domestic dog are referred, has never been conclusively established. But three groups of ancestors have been suggested and all of them are plausible: wolves (*Canis*

[1] Hyams (1971).

lupus), jackals (*Canis aureus*) and various wild dogs such as the pariah, the dingo, the pi-dog and, above all, a species now known only as a fossil, the *Canis poutiatini* of Russia. It has also been suggested that our domestic dogs may descend from hybrids between two or more of these groups, and all the possible combinations of ancestry have been put forward by various zoologists who often disagree among themselves.

Considering the very great differences in size and shape of the almost innumerable breeds of domestic dog, it seems likely that the lines of evolution in domestication must lead back to more than a single origin; and that the crossing of the lines, from time to time during the enormously long history of the association between man and dog, has complicated the whole story.

Studies by experienced authorities of the bones of both dogs and wolves taken from Maglemosian sites in north-western Europe have shown one thing quite clearly: that the Maglemosian dog was indeed derived from the wolf, and that it could certainly cross-breed with wild wolves. It is to the point here that wolves are easily tamed if caught young (see page 8), and also that all the elements of the social behaviour of dogs can be traced back to the social behaviour of wolves. The wolf here in question is the small-sized race inhabiting south-west Asia, whose range was formerly very much greater than it is now. So, although there is no suggestion that all dogs descend from this little wolf, it does seem clear that some of the larger breeds do.[1]

So the Maglemosian dog was a sort of wolf. This Maglemosian culture was a Mesolithic one, falling between Old Stone Age and New Stone Age, and, in the region in question, between about 6000 B.C. and 1500 B.C.; it was a north European culture.

Next there is the theory of the jackal as ancestral to the dog. One of its principal exponents was Lorenz, who argued

[1] Reed (1960).

that certain breeds of dog, which he calls 'aureous dogs', behave like jackals; that wolves and jackals can interbreed; that 'aureous dogs' and 'lupus dogs' would have interbred; that jackals are easy to tame, and readily associate with man. But having long supported this theory, Konrad Lorenz himself turned against it and abandoned it.[1] One argument against the theory is that the young of a dog-jackal cross are sterile, This, however, is by no means established, and excellent authorities still accept the jackal as one of the ancestors of the dog.

What is very generally accepted as the earliest race of domesticated dogs, earlier even than the north-west European Maglemosian dog, is the one whose skull was found on Natufian sites in excavations on Mount Carmel in Palestine. This animal was clearly derived from a jackal. In fact, the resemblance between this skull and the jackal's skull was found to be so close that Reed (1960) suggested a very simple explanation, that this was no dog, but a jackal indeed. Perhaps; but then, is not a jackal a dog, and are not jackals easily tamed? Still, here is the crux of his argument:

Romer (1938) shows how confusion between bones of dogs and jackals may lead to erroneous ideas as to the antiquity of the dog, which had been reported from several 'palaeolithic' sites in Algeria. He believes that all such instances were due to hopeful misidentifications of jackal bones. His findings should be a warning to all investigators who apply the word 'domestic' to any familiar-looking bones.

Well, I suppose it depends on what you mean by 'dog'; a domesticated jackal is a sort of dog and not by any means out of the question.

Then there is the theory of domestic dogs as descended from wild dogs, if there really and definitely is such a thing as a wild dog (by which I mean that the animals may be wild, all right, but derive from domestic dogs gone feral).

[1] In a verbal communication to the Ethnological Conference at Cambridge in September 1959.

Here it is argued that when we compare the skeletons of dogs found on Neolithic sites with those of wolves, the morphological differences are so great, notably in the case of the skulls, that the wolf cannot have been ancestral to those dogs. This does not, I believe, mean that, for example, the Maglemosian domestic dogs cannot after all have been wolf-dogs; it does, surely, imply more than one domestication, more than one origin. On the positive side, the argument is that these Neolithic dogs descend from an animal now extinct and known only from bones, *Canis poutiatini*, associated with Palaeolithic cultures. This implies that hunting people in what is now Siberia or Russia had domesticated a dog, or at least had some kind of hunting partnership with a dog, about 10,000 years ago, or more. Living races of feral dogs (dingos, pi-dogs and pariahs and some others) are thought to have descended from this same wild dog or very early domestic dogs—surely a remarkable instance of wide diffusion, if true.

The wild dog theory in one variant or another has had much support. Some of this would seem to be emotional, on the part of dog-lovers who abhor wolves and whose attitude is fostered by the survival in our culture of Medieval European folk-tales of the wolf. However, the theory has several serious supporters (see Sauer 1952; Vesey-Fitzgerald 1957; Bodenheimer 1958; *Illustrated London News* 1958). As Bodenheimer remarks, what is needed is a thorough study of the neglected pariah, still found throughout much of the Eastern Hemisphere, but almost unknown zoologically because of the general assumption that it is a half-feral domestic dog.[1]

The earliest remains of unquestionably doggy canids yet found come from Jericho, from what archaeologists call the 'plaster-floor level' in the excavation at Tell al-Sultan: these seem to have been wolf-dogs, and there were two different breeds. Then, the earliest occurrence of an identifiable breed,

[1] Reed (1960).

a breed still with us, is that of the saluki. This was a Mesopotamian domestic dog; it began its career in the Ubaid
period (*before* 3000 B.C.) and was a heavier animal than the
modern saluki. It was presumably a hunting dog. Somewhat
later appears the first Egyptian domesticated dog. I say
later, but how do we know when? In this case the evidence
is a painting, and there is no reason to suppose that as soon
as the Egyptians had domestic dogs somebody drew their
picture. However, this earliest representation is a painting
on a pottery bowl, known as the 'Golenischeff' bowl, which
belongs to the Amratian period (*c.* 3000 B.C.). The picture
shows four ferocious-looking dogs held in leash by one man.
The dogs could be salukis, but are rather more like greyhounds.[1]

Skeletal remains of pre-Dynastic dogs from Maadi and
Heliopolis have been shown to be much more wolf-like than
jackal-like. Now the true wolf, *Canis lupus*, is not and never
has been native to any part of Africa; so the earliest Egyptian
domestic dogs were introduced, presumably by way of
Palestine, from Mesopotamia, in which connection the two
dates cited above are interesting. Incidentally, Pierre
Montet (1968) claims that the pre-Dynastic Egyptians may
have domesticated the hyaena as a hunting animal long before
its domestication as a source of meat; could this have been
before the dog in Egypt?[2]

Were there numerous domestications of the dog—or perhaps one should say *creations*—with the wolf, the jackal and
the wild dog as raw material? Or was the earliest domesticated
dog disseminated from one centre, to develop, with enormous
differences, into local races? Egypt does seem to have had
her first dogs from Mesopotamia, but it does not follow that
there were no other sources, and it will be as well to glance
at some other parts of the world.

[1] Massoulard (1949), Plate XXXII.
[2] Montet does not mention the use and possible domestication
of the striped hyaena in the Fifth Dynasty—for food.

In Siberia, where the Magdalenian (Upper Palaeolithic, say, *c.* 35,000–*c.* 9000 B.C.) culture phase, entailing the hunting of big game for a living, lasted later than in Europe for reasons which seem to have been climatic—the persistence there, later than farther south and west, of the flora and fauna constituting the environment in which the Magdalenian culture developed—the dog was domesticated very early. According to Vadim Elisseeff,[1] basing himself on finds from excavation at Afontava Gora in Siberia, man and dog were already associated in a hunting partnership of some kind in that part of the world more than 10,000 years before our era. Both dog and man hunted big game in packs, using hunting tactics directed by leaders; perhaps, instead of competing, they learnt to co-operate. This, of course, puts the domestication (or something like it) of the dog—Elisseeff claims it as the first in the world—very much earlier than those we have discussed above. But again the question arises, what is a dog? Maybe the animal was a wolf. But there is at least one reason to think that some kind of canid had been domesticated very early indeed in Siberia: it was from Siberia that men peopled the Americas, and it does seem as if some kind of hunting dog went with an early wave of (Palaeolithic) migration.

A few other items of evidence for dogs in remote antiquity: there is some pictorial evidence for the use of guard-dogs to herd cattle in pre-Dynastic Egypt. How did the dog develop from a hunter and killer to a protector and marshal of cattle? In precisely the same way and at the same time as man himself: when game is plentiful, and in view of the fact that, if you then kill too much, most of it will go bad before you can eat it, there is good sense in driving, corralling and keeping your meat on the hoof. And by the time man had thought of that one, the dog had been so long associated with him that it could be taught the same tricks.

In China the dog makes a first appearance in the Neolithic period. I cannot find an authority who commits himself to a

[1] In Varagnac (1968).

date for the excavations at Pan-p'o-ts'wen in Shensi, but it
is apparently about 2500 B.C.[1] The excavations exposed an
entire Neolithic village whose inhabitants, although farmers,
were still, also, hunters. They had flocks of sheep or goats,
herds of swine and, to help them guard those animals as
well as to help them in their hunting, a domesticated dog.
In Japan the dog makes its appearance in human communities
long before any other domesticated animals, in the Neolithic
Jomon Culture. This suggests an introduction from China;
and the Chinese dogs may have come from Siberia.

In England the domestic dog makes its first appearance
with the Neolithic Windmill Hill people, immigrants from
France, pastoralists with herds of large, long-horned cattle
for which they built big corrals by digging ditches. They
were the Long Barrow people; and with them they brought
a dog which doubtless helped the herdsmen and guarded
the herd against wolves.[2]

Having established as best we can roughly where, when
and in what conditions the dog was first domesticated—
first perhaps in Siberia in a Magdalenian context, but possibly
in similar contexts elsewhere and certainly in a number of
places in a Mesolithic context, so that by Neolithic times
dog and man had long been friends—we have to consider
just how this remarkable partnership came about.

Wolves, jackals, in fact all wild dogs, are scavengers and
carrion-eaters. When a band of human hunters had killed
a big-game animal, or, as sometimes happened, had cut out,
corralled and slaughtered hundreds of animals of a great
herd, wild dogs would soon have been on the scene in the
hope of seizing their share of the meat. They would soon
have learnt, too, that they would not always be driven off;
when game was plentiful there were probably parts of the
animal which the men did not eat, some of the offal perhaps.
So encounters between hunting packs of men and hunting

[1] Elisseeff, in Varagnac (1968).
[2] Hawkes, J. and C. (1944).

packs of dogs must have been frequent; and although this might not, since they were in competition, promote friend-ship between the two species, the dogs or wolves or jackals—let us call all of them dogs—would have learnt that there was food to be had where men, much cleverer hunters than them-selves, were driving game; and about men's camps. Jackals and young wolves are very easy to tame, and none of the dog family is afraid to enter the domain of man. On the other hand, the smaller kinds, all but the most ferocious of the larger wolves, are not a serious danger to man and are afraid of him at least to the point of being, as a rule, inhibited from attacking him. Thus, in India wild jackals are valued as natural scavengers of food debris, and are accustomed to roam not only rural settlements and villages, but even large towns in search of food.

In short, wild dogs of all sorts would have been accustomed to visit man's encampments, and in certain circumstances would have been tolerated and even welcomed.

Now, Zeuner has a most interesting suggestion as to the origin of the domestication of dogs, one that is, to my mind, pleasing and even rather touching. Having discussed some cases of creatures of different genera living together in close association, only less close than those symbioses achieved by the lower forms of life, he goes on to show that social animals have a marked tendency to support alien forms of life which, albeit intrusive, are not openly hostile or dangerous. One recalls how people will persist in feeding such vermin as sparrows and mice. During the Second World War I came across a case of British soldiers deliberately feeding rats which infested the building where they were billeted, until the rats became a danger to small children of the local resi-dents and the practice of feeding them had to be stopped by an order from the Commanding Officer. People are on record as keeping every conceivable kind of beast, even crocodiles, as familiar pets. Zeuner (1963) says there may well have been a stage of throwing morsels of food to wild dogs which invaded the camps of pre-Neolithic man, long prior to any realization of the economic use of dogs:

Such acts are elementary manifestations of the solidarity of life, especially of related life. It is characteristic of most higher animals which have developed a social medium of some sort and which are not enemies. It finds a simple expression in animal friendships as they occur under conditions of domestication, between cats and dogs, or stranger still between cats and tame birds.

Gilbert White (1789) has accounts of several such friendships, one in particular between a horse and a hen, which would seem wildly unlikely had not so good a naturalist observed and recorded it. Nor, returning to men and wild dogs, can one completely ignore the great number of 'Mowgli' stories—tales of children reared by wolves and running with the wolf-pack; Romulus and Remus were only the two most distinguished of scores of wolf-children.

The next stage is pet-keeping. Here Zeuner suggests that this tendency to suffer the presence of alien species as pets may begin very early, in a Palaeolithic context. It seems that Australian aborigines capture and tie up dingo puppies until they are tame, after which, untied, they will not run away. Here is a pleasant tale from Nigeria (Calabar) which, though legendary, may reflect the very act, remote in time, of dog domestication or something very like it. A village boy caught and adopted a wild-dog pup and became much attached to it; he brought it up in the village, despite its mother's attempt to recover her puppy. When this puppy had grown up to be an adult dog, it persuaded a wild bitch to come and live with it in the village. Thus, their litters were born in domestication and, having grown up, took to going out on expeditions with the hunters who were now their friends. Neighbouring villages saw the advantages and imitated it, and so the dog became a domestic animal.[1]

[1] Lane (1946).

As I have said, the smaller kinds of wild dogs are afraid of man but can overcome their fear to the extent of approaching him; they are sociable; they can live well on the parts of meat which men esteem least, and on other food debris; they can even manage to turn vegetarian; and as they are accustomed to run in packs led by a recognized leader, they have the instinct which enables them, by transfer, to accept man as a pack-leader if he be clever and strong enough to deserve the post. Finally, another quotation from Zeuner (1963):

Assuming that the presence of wild dogs was suffered in the camps of Mesolithic man, because of their usefulness as scavengers, the habits of dogs afford a possible explanation of the development of closer association between dog and man. All wild dogs exhibit the beginnings of a social life, in so far as they associate with each other to form packs for the purpose of hunting as well as for defence. The individuals of each pack are tied together by bonds of friendship, they help one another and often, in their operations, work in unison according to what one might almost call plans designed to meet certain situations. Their intelligence is far superior to that of other carnivores. . . . The cunning of the dog and his concerted efforts on hunting expeditions are not unlike the hunting practices of early man. It is therefore not impossible that young wolves or jackals or other wild dogs, which grew up in or near the temporary camps of Mesolithic hunters, would quite naturally regard the men, who provided part of their food supply, as members of their pack, an association which the hunters would not have failed to turn to their advantage.

2

Asses, Onagers and Horses

There are three principal species of the genus *Equus*, although many more specific names will be found in the numerous authorities. These three wild equids have been used by man in the creation of all the races and strains of working asses, onagers, horses and mules. All are confined in nature to the Old World: although horses were ultimately to become feral on the great plains of North America, and on the *pampas* in the temperate zone of South America, no horse had been seen in the Americas until brought by the Spaniards. The equid which according to fossil evidence evolved in America in the Eocene became extinct before man's arrival in that continent. We therefore begin this enquiry with at least one certainty: asses, onagers and horses were domesticated in the Old World.

The horses called *caballine* (from *Equus caballus*) were originally native to the whole of Central Asia and a part of Eastern Europe north of the great mountain ranges; that is, roughly to the vast European steppe. The original home of the ass, *Equus asinus asinus*, now extinct as a wild animal in two of its races and not quite extinct in the third, was North Africa. The onagers, *Equus hemionus*, were distributed in a number of geographical races throughout Arabia, Iran, Turkestan, Afghanistan and India. A fourth principal species of horse, consisting of the several sub-species or geographical races of zebras, has a vast natural range in Africa, has never been domesticated and therefore does not concern us here.

Three geographical races of *Equus caballus* survived in the wild state almost until modern times. The race called

Przewalski's Horse is perhaps even now not quite extinct in the wild; if it is, then extinction must be of recent date since there are specimens of the race in zoos, and small herds of it have been built up by naturalists. A race known as the Forest Wild Horse was still represented by living wild horses in some Polish forests until the middle of the eighteenth century. The Tarpan, the most important of these races from our point of view, survived in the wild state into the nineteenth century, and it is at least not impossible that somewhere in its vast range of steppe and foothill a small herd still persists.

We have, then, three large territories where equine animals could have been domesticated: Eurasia, Central Asia and eastward into China (Tarpans in the west, Przewalskis in the east), and North Africa. It is also possible that there were wild horses in Spain in prehistoric times, and there are reports of herds of wild horses in the Alps and on the Italian plains in the eighth century A.D.[1] We can also limit the temporal range within which domestication occurred. There are no reliable traces of horse domestication until Neolithic times; horses were not domesticated until man had become a farmer.[2]

The true horse was the last of the equine animals to be domesticated. It is not absolutely sure which came first, the ass or the onager. It is curious, however, that, despite the contrary assertion of ancient Roman authorities who knew what they were talking about,[3] the onager has long been supposed to be untamable; it was, and is, nothing of the kind.

Although the onager, the Hemione, of Mesopotamia is now extinct in that country, it has only quite recently become so. Layard, pioneer of archaeological excavation in Meso-

[1] Heyn (1888).
[2] Drower, in Ucko & Dimbleby (1969).
[3] Pliny and Columella. The latter says (VI.37) that onagers could be crossed with asses. They may also have been crossed with horses to produce mules, but not until after 1000 B.C.

potamia, saw herds of wild Hemione between Tigris and Euphrates after 1845.

When Sir Leonard Woolley excavated the Royal Cemetery at Ur of the Chaldees, he found, among other treasures dating from *c.* 2500 B.C., some pictures of rather mule-like animals drawing a chariot.[1] His claim that the draught animals in question were onagers was rejected (owing to that nonsense about the onager being untamable), until a large number of bones excavated at Tell Asmar were identified as those of onagers. Later, more representations of domesticated onagers came to light: an electrum figurine of an onager decorating Queen Shub-Ad's chariot pole; a rein-ring bearing the copper figure of an onager; and on the Royal Standard of Ur. There is even a possibly much earlier record: a fragment of a vase excavated at Khafajeh bears a picture of mule-like animals, which may be onagers, pulling a cart. This fragment from the Jamdat Nasr period is dated *c.* 3000 B.C.

Zeuner (1963) makes a cautious reservation about the apparent domestication of onagers in the third millennium B.C. or even (if the dating of the Jamdat Nasr fragment be correct) late in the fourth. He says it is not certain that at that early date the onager was domesticated in the sense that its breeding was controlled by man, and that the Sumerians may have obtained their animals by catching wild foals. True, there was found in the Palace of Ashurbanipal at Nineveh a beautifully clear and lively representation of onagers being captured with some kind of lassoo; [2] but I see no reason why both capture and breeding should not have been in use at the same time. Nor do I know any reason to assert that the onagers in that representation were necessarily wild; domesticated but still unbroken foals on horse ranches often have to be captured with lassoos, and for all we know the two men shown leading the roped foal in the Nineveh picture were capturing it not from a wild but from a tame

[1] Woolley (1934).
[2] It can be seen in Room S. B.M. No. 124882.

herd out at grass. They were probably engaged in breaking it in. I think we can take it that the onager was domesticated by the Sumerians in the third millennium B.C.

There were formerly three geographical races of the asses: a North-east African race which was at home in the mountains of Nubia and in the eastern Sudan—it is now either extinct or very nearly so; a North-west African race which became extinct in Roman times or soon afterwards; and a Somali race. This last survives and is, happily, protected. The two North African races were the ones that gave rise to the domesticated asses. There were also European and Crimean races of this animal (quite distinct from the onagers) in prehistoric times, but it seems that they became extinct long before any domestication was conceivable; perhaps they were hunted to extinction, for meat. It is a great mistake to imagine that our generation alone is guilty of exterminating beautiful species of animals and birds; mankind has, unfortunately, been doing just that for thousands of years. At all events, we need not look beyond North Africa for the place where asses were domesticated.

As we should expect in that part of the world, the Egyptians or the Libyans were the pioneers. Some authorities have opted for Libya (not in the past as arid as it is now, for it is largely a man-made desert) on the grounds that asses are shown on the Nagada Palette as one of the elements of tribute from Libya to Egypt. Libya or Egypt, we have a date; for a panel from the Fifth-Dynasty tomb of Sahure—1650 B.C.—depicts asses among the domestic animals.[1] As I have said elsewhere, such dating is useful only inasmuch as it enables us to say that domestication of the animal in question was not later than the date of the representation: not only could there have been earlier representations which we have not yet found and perhaps never shall, but also, of course, there is no reason why the animals in question should have been drawn by an artist as soon as they had been domesti-

[1] Zeuner (1963).

cated. There was, alas, no *genre* painting in those days;
or, if there was, it has not survived. So all we can say is that
the domestication of the ass was accomplished by the Egyptians or the Libyans in the third millennium B.C., conceivably late in the fourth (although, on other grounds, the earlier
date is unlikely).

From Egypt the domesticated ass was introduced into
Syria and Palestine: the cuneiform name of Damascus reads
literally 'town of asses'. The ass reached Mesopotamia in
the second millennium B.C.; it was used by the Hittites.
How did this come about? Probably by the movement of
rich nomadic Bedouin sheikhs—princes of the desert and
the steppe, like Abraham—between the two centres of urban
civilization.

When Abram arrived in Egypt, the Egyptians saw that she
[Sarai] was indeed very beautiful. Pharaoh's courtiers saw
her and praised her to Pharaoh, and she was taken into
Pharaoh's household. He treated Abram well because of her,
and Abram came to possess sheep and cattle and asses, male
and female slaves, and she-asses and camels. . . .[1]

An unedifying period in the life of the patriarch, but he was
not the first or last man to grow rich by exploiting his wife's
sex-appeal for the man who could make his fortune. What
was the use of those asses and she-asses? The asses were
used as beasts of burden; as for the she-asses, the ancient
Jews were, together with the Nubians, probably the first
people to ride them. For until the Middle Kingdom the
Egyptians used them simply as beasts of burden, later for
dragging their ploughs and for treading out the corn.

The ass is indisputably one of the most useful animals, yet it is
despised nearly everywhere. It is not fully understood why
this should be so. In part, its stolid temperament has annoyed
its master since time immemorial. By comparison with the
horse and mule it is inferior. Its food requirements are
extremely modest, thistles and straw are sufficient, a diet

[1] *New English Bible.* Genesis, 12.

with which the horse would not be satisfied. Modesty, however, was not in ancient times a quality of character regarded as worthy of admiration. In civilized countries the ass became the beast of the poor; of those who could not afford to buy and maintain the more pretentious animals. The patience of the ass likened it to a slave. It is probable that all these factors contributed to its low position in the social scale of domestic animals.[1]

So true is all this that not even the fact that Jesus entered Jerusalem riding on an ass did much to improve the poor donkey's social standing. Dr Zeuner, however, continues:

Originally the ass was not despised. The Egyptians were proud of their white asses, and the large and graceful asses of Muscat have until recently served as processional mounts for the family of the Sultan of Muscat. Some Romans, too, appreciated the beauty of the well-bred asses. Varro relates that the Senator Q. Axius paid 400,000 sesterces for two pairs of draught asses, and Pliny that an ass from Reate in Sabinia was sold for 60,000 sesterces. This province was renowned for its excellent asses, as was Arcadia in Greece. Probably because of its association with corn and flour, the ass was the sacred beast of the goddess Vesta. In 260 B.C. a Cn. Cornelius Scipio Asina was Consul, and his name is believed to be due to his large ears. If so one might assume that at that time the ass had not yet been reduced as much in estimation as in later times.

Zeuner's conclusion seems to me excessively kind. I know no reason why the lieges of 260 B.C. should have had any more illusions about politicians than the lieges of 1971. It is curious that whereas Hesiod's *Works and Days* never mentions the ass at all, that it is not mentioned once in the *Odyssey* and only once, by way of a simile, in the *Iliad*, mules are mentioned more than once in all three works. The presence of the mule implies that of the ass, but Greece appears in Homer's time to have imported mules from the Henetians, a Paphlagonian people of the Pontus:

[1] Zeuner (1963).

Where rich Henetia breeds her savage mules

(Iliad).

And mules were evidently acceptable as tribute:

Last to the yoke the well-matched mules they bring,
The gift of Mysia to the Trojan king.

(Iliad).

During a period of more than 4,000 years following its domestication, the ass's milk became famous both as medicine and as cosmetic, its dung as a medicament and the best manure for pomegranates, its skin as parchment, while the animal itself became the divine avatar of King Midas, a symbol of sexual potency for Roman newly-weds, and in Egypt, a god who suffered crucifixion in the third century A.D.[1] During those forty centuries, too, the ass was confined to the Mediterranean, approximately to the same range as the wine vine and the olive; it did not become common and familiar in northern Europe until the early Middle Ages— very much later than the horse.

The horse made its first European appearance in Macedonia, during the Bronze Age; but there is as yet nothing to show whether it was domesticated or wild. The earliest European representation of a horse clearly domesticated, for it is depicted drawing a two-wheeled chariot, is Mycenaean and dated to c. 1550 B.C. By 1300 B.C. at the latest, probably earlier, the Mycenaean Greeks not only had horses but were also riding them.[2] Doubtless the Mycenaeans had their horses and chariots from their Cretan cousins; at all events, the horse and chariot are a little earlier in Crete. And I suppose that the Cretans, in their turn, had horse and chariot from the west Asian mainland; they were in use there by 1800 B.C. Dating by representations is necessarily conservative:

[1] The asinine crucifix of Typhon-Seth was found on the Palatine hill in Rome.
[2] Terracotta model of horse with rider, Lower Helladic III b. In the Metropolitan Museum, New York.

one can safely say that the horse was well established in Greece by 1600 B.C., but was probably a rare novelty in 1700 B.C.[1]

It will be useful now to take a look at the earliest records of the horse in the advanced centres of urban Bronze Age culture in western Asia and on the Nile. The first written notice of what were probably domesticated horses is in a document belonging to the Third Dynasty of Ur of the Chaldees, c. 2100 B.C. The animals are referred to by a name which apparently means 'foreign ass'. In a Sumerian story entitled *Enmerka and the Lord of Aratta*, whereas asses are used as beasts of burden, these 'foreign asses' (ANSE KUR. RA) are not; and since the horse in antiquity is distinguished by, among other things, *not* being used as a beast of burden, it may be concluded that these 'foreign asses' were horses.[2] They must have been not only foreign but also rare, perhaps seen only occasionally; for the horse is not mentioned in the Code of Hammurabi, c. 1800 B.C., and as this first great code of laws dealt with every aspect of the social and commercial life of the Babylonian Empire, it is clear that the horse, if known at all, was not in common use.

A little later the horse begins to appear as an article of rare and costly trade between the Amorite princes, with Carchemish on the Euphrates as the major horse-market and Cappadocian Kharsamma as the source of supply. It also appears in Accadian texts in the eighteenth century B.C.; the name for it is *sisu*, a word that seems to have an Aryan root.[3] At Brak and Chagar Bazar on the river Khabut, two other eighteenth-century B.C. sites, models of horses with painted bridles were found, and also some tablets referring to 'yokes' of horses, which implies harnessing for draught. The point to note is that the Bronze Age horse appears in the less advanced centres at about the same time as in the

[1] Zeuner (1963).
[2] Salonen (1956).
[3] Salonen (1956).

more advanced, and perhaps even a little earlier. The reason
for that would have been geographic rather than cultural.

The vehicle to which those horses were harnessed was
the first horse-drawn chariot; that is, the horse was used
in war for centuries before any other, any really economic,
use was made of it. This chariot has a clear ancestry: it was
probably the second or third generation of wheeled vehicle.
Some time before 3500 B.C. the people of the Uruk Culture
in southern Mesopotamia invented a wheeled cart, together
with the potter's wheel. Before that time carts moved not
on wheels but on runners; they were sledges.[1] These, and
probably also the earliest wheeled carts, were drawn by oxen.
In due course the ox was replaced by the much faster onager,
which had even been used to draw sledges, at least those of
royalty. Gordon Childe established that wheeled vehicles
reached the Indus Valley—Mohenjo-daro and Harappa—
c. 2500 B.C., Syria c. 2200 B.C., Crete three centuries later,
Greece in the middle of the sixteenth century B.C., Italy
two centuries later, Continental Europe c. 1000 B.C. and
Britain c. 500 B.C. This new technological device evolved as
it spread across the Eurasian world; horses and vehicle probably arrived in each new outpost of civilization at about the
same time, very often at exactly the same time in the form
of an attack or invasion by chariotry.

What sort of people were the charioteers? It will be
appropriate here to observe that from the very beginning of
its association with man the horse was 'noble', and usually
more or less sacred. Its use was long confined to war, and
always to princes and nobles; to be mounted was to be
superior, and that was the origin of chivalry (*caballarii* =
chevalerie = chivalry). The same social phenomenon occurs
everywhere: in China, for example, and Japan, as well as in
western Asia, Arabia, Africa and Europe. Were not the
Argentine gauchos a kind of chivalry? Whoever has read
Hernandez's *Martin Fierro* cannot doubt it. It is not too

[1] Childe (1951).

much to say that the tamed horse created social nobility. No wonder democratic mobs resent mounted policemen: they make them feel as socially as they are economically inferior.

It was this semi-sacred nobility of the horse which gave rise to the very widespread tabu against eating horse-flesh. It is a tabu which has nothing whatever to do with sentimental regard for the horse as the friend of man: you do not eat or allow to be eaten your totem animal, except ritually on rare occasions. It must have arisen in the Bronze Age with the arts of driving and riding; for horses were certainly eaten in Neolithic times and, as will appear, the origin of the domestication of the horse is to be found in this very practice of hunting and eating it. The tabu was, and remains in some places, for example in England, very persistent and hard to break. It had the powerful sanction of the early mediaeval Church. Thus in 732 Pope Gregory III wrote to St Boniface: 'Thou hast permitted to some the flesh of wild horses and to most that of the tame. Henceforward, holy brother, thou shalt in no wise allow it.' [1]

The horse makes its first appearance in Egyptian records, together with the war chariot, in the sixteenth century B.C.; at all events, that is the approximate date of the earliest representation we have found—an Eighteenth-Dynasty one, c. 1570 B.C. But apparently the horse was in use a little earlier than that: it is mentioned as being used by the army, in texts concerning the war of liberation against the Hyksos. At about the same time, horses and chariots were taken into use in India.[2] The earliest Chinese culture to have any knowledge of the horse was pre-Bronze Age—the Neolithic Lung-chan, in Hupei, 2000–1500 B.C. But as to the manner of use, there is no archaeological evidence; quite likely the horses were kept for meat, being simply corralled rather than domesticated.

[1] Heyn (1888).
[2] Drower, in Ucko & Dimbleby (1969).

Przewalski's Horse was native to China, but it seems not to have been domesticated until after the idea of domesticating the horse, and perhaps the domesticated horse itself, had been introduced from a source west of China.[1] By the end of the Shang Dynasty the horse and chariot were in use, as in Egypt, western Asia, India and Mycenaean Greece. The Chinese may have been the first urban, civilized people to use mounted archers; and it was as the use of horses in war became familiar to them that they took their own wild horses into domestication. But these, being Przewalskis, were never as good as the horses of more westerly peoples; this is clear from the fact that in 128 B.C.[2] the Emperor Wu of the Han Dynasty sent General Chan K'ien to Ferghana, Sogdiana and Bactria to buy horses for the Chinese army. Those Iranian horses were of Tarpan stock and much superior, for it is the Tarpan of the Eurasian steppe that has given rise to all the noblest domesticated races of horse.

Chan K'ien's expedition sent home some 3,000 Iranian horses.[3] Chan K'ien was evidently a most efficient officer, for together with the horses he sent seeds of *Medicago sativa*, alfalfa, a crop plant greatly superior to anything which the Chinese had at home, and thus introduced an invaluable fodder plant to his country.[4]

The picture which has now emerged is, I hope, fairly clear: between *c.* 2000 B.C. and *c.* 1300 B.C. the advanced, urban, Bronze Age cultures of the Nile Basin, Mesopotamia, Syria, India, Greece and China receive the domesticated horse from a source lying between them, and to the north. They all, within a relatively short time, begin to use a light, horse-drawn war chariot. With the horse come the skills of

---

[1] Zeuner (1963).

[2] Laufer (1919). But Zeuner (1963) says 102 B.C.

[3] Also some mysterious animals called Heavenly Horses for the Son-of-Heaven's eventual funeral procession. We know nothing about them except that they 'sweated blood', which Zeuner (1963) explains by reference to a possible skin-wounding parasite.

[4] Laufer (1919).

stable-boy, groom and horse-leech; for throughout Mesopotamia, Syria and Anatolia the evidence from cuneiform
tablets is that with these horses from the north came breeding techniques and the skills of horse-breaking and veterinary
care. Drower quotes a letter from a Hittite king to the King
of Babylonia in which the writer asks for a new supply of
horses because those sent to him by a former Kassite king
have all gone lame: 'In Hatti-land the cold is severe, and an
old horse does not last long. So send me, my brother, young
stallions.' Such stallions are, in Kassite texts, given names—
'Foxy', 'Starry', 'White', 'Piebald'. And since the name
of the animal's sire is often mentioned as well, Drower [1]
suggests that stud books may have been kept; at least the
notion of pedigree was already understood.

That all those horses were not much more than pony-
sized is clear from pictures of noble warriors driving the
monoplace chariot, from skeletons and from the dimensions
of the stables at Ugarit. They were always driven in harness,
never mounted (unless occasionally by a groom); but by
the fourteenth century B.C., at latest, horses were also being
ridden—bridled but not bitted, and bare-backed, if a drawing
from the tomb of Horemheb at Memphis is reliable. Riding
in general had to wait for the invention of the bit (not a long
time, for the bit was a Bronze Age invention), the saddle
and, above all, stirrups.

As always, in the case of domesticated animals and plants,
local races were developed. Once breeding was understood,
horses could be bred for speed or size, or for weight as larger
chariots were developed, or chariots which had to be armour-
plated to withstand improving projectile and missile weapons
such as the double bow. A theory that there must have been
a separate and different domestication of the horse in Europe
has been advanced to explain the huge size of the horses
required to do heavy farm work, or to carry armoured knights.
But there is no real need for such a theory; selective breeding

[1] In Ucko & Dimbleby (1969).

for size, as the work became heavier, will account for those massive animals.

Here we had better go back to that linguistic feature which indicates the ancient Aryan homeland as the horse's point of departure. The place must have been somewhere on the great Asian steppe where vast herds of Tarpans roamed and grazed. But how did the partnership of man and horse begin?

When one considers the domestication of any kind of cattle, a pattern becomes clear. People who have for thousands of years hunted a particular species for meat realize that the best way to preserve meat in good condition, when game is plentiful and some provision can be made for a leaner time ahead, is 'on the hoof'. They discover also that with some species this is possible. Palaeolithic man could not corral a herd of elephants, but he could corral sheep. I have no doubt that in some places the practice of driving, corralling and keeping herds of beasts developed. This pattern could still be studied in what one may almost call historic times, despite the absence of writing, in Inca Peru. There, while some of the animals belonging to the genus which includes Llama, Vicuña and Alpaca were truly domesticated, great herds of the wild ones, living at very high altitudes, were guarded but not domesticated. From time to time there would be an officially organized mass hunt: part of a great herd would be cut out and driven into prepared pens, or kraals; then some of the young were taken for domestication, all the animals were sheared, and some kept penned for slaughter.

In Palaeolithic and Mesolithic times the horse was hunted for meat in Central Asia; I have no doubt that late in the Palaeolithic men had learned to corral parts of a horse herd and keep it on the hoof until wanted for eating. This would probably, due to the tractability of the foals, have developed in the Neolithic into true domestication: the guarding of the herd, its breeding in captivity, the rearing of its young, the protecting of it against carnivorous predators. So at first the community of men lived parasitically on a herd of horses; but then the relationship developed into something more subtle and complex, something which it is hardly stretching the word too much to call symbiotic.

Thus, tame herds of meat-horses and perhaps milch-mares were there for the harnessing when the horse people conceived the idea of using animals for their muscle-power. At the periphery of their steppe advanced Neolithic people were harnessing oxen to drag freight and even passenger sledges; onagers, too, had been broken to harness. So why not the horse? Surely there can be no doubt that the first drivers and riders of horses were people who had been living on mare's milk and horse meat, and who received the idea of bridling and harnessing horses from neighbours who had already begun to use other animals for draught.

In the Ukrainian Tripol'ye culture, as revealed by archaeological digs on a number of sites, the bones of horses are found at all levels, but in smaller numbers in the lower, most ancient levels than in the advanced, higher levels, at which point they become extremely numerous. The most obvious and plausible explanation of the increase is domestication at a time represented by a level between the two extremes: 2800 B.C. is a reasonable guess; a thousand years later there were ceremonial horse-burials in what is called the Timber Grave Culture of the Volga region.[1] The conclusion is, then, that the horse was first domesticated from wild Tarpans on the Eurasian steppe earlier rather than later in the third millennium B.C., and perhaps first broken to harness a century or so later.

We have seen that the horse was driven before it was ridden. Riding, as a general practice, not an occasional stunt by stable boys and grooms, came later, with the invention of bit and stirrups. The second great wave of migration of Aryans from Central Asia into Iran occurred about 1000 B.C. This time the hordes were mounted, a fact apparent from finds by archaeologists of horse-furniture and models of mounted

[1] Clark & Piggott (1965).

horses. The shock to the peoples of the invaded lands must
have been paralysing: their foot soldiers and chariotry were
no match for those terrible cavalrymen who had learnt to
shoot from the saddle and whose swiftness in manœuvre
was bewildering; one recalls the terrifying impression made
by mounted Spaniards on the Aztecs of Mexico and the
people of the Inca Empire in Peru. True, west Asians,
Europeans and North Africans were familiar with the horse
and had occasionally seen horses ridden; they were not apt
to make the mistake made by the American Indians, who at
first thought man and beast constituted a single animal—
or god. All the same, massed cavalry was as new in warfare
then as tanks were in 1917, and as devastating to infantry.
Moreover, the horses of that second wave of invaders were
larger than the familiar chariot horses; doubtless they were
the product of deliberate breeding for size.

The lands south of Lake Urnia became a great centre for
horse-breeding, and it was from there that the Assyrians
obtained their horses. Those Assyrians were certainly the
first urban, civilized people to have regular corps of cavalry,
which made them the most formidable militarists of their
time, able to dominate their century (900 B.C.).

That is not the end of the story: there were peoples who,
as a result of the Tripol'ye domestication of the horse, created
remarkable horse-cultures. The first and most interesting
among them, of whom we have historical knowledge, were
the Scythians; but it is at least possible that they had learnt
the elements of their way of living on, by and with horses
from the Huns, who, time after time, were to shake the stability of the sedentary farming-based civilizations to the west
and east of them by means of their complete mastery of the
horse and of their ability to live on nothing but grass by
exploiting the horse.

Herodotus describes how, in the fifth century B.C., the
mares of the Scythians were milked by blinded slaves who
had been taken in war; how the milk was 'stirred around'

to make koumiss; how the Scyths sometimes made wholesale
sacrifice of horses to their gods. As to where they learnt this
way of life, Herodotus says:

There is another and different story [touching the origins of
the Scythians] . . . now to be related which I am more inclined
to put faith in than in any other. It is that the wandering
Scythians once dwelt in Asia and there warred with the
Massagetae but with ill success; they therefore quitted their
homes, crossed the Araxes [Volga] and entered the land of
Cimmeria [Crimea]. . . .

The Massagetae were the Huns, a people better adapted
to living on horses than any other in the world's history;
not even the gauchos of the Argentine *pampa secca* created
quite so perfect a horse-man symbiosis. Perhaps the Huns
were the folk who taught the Scythians their horse-lore.

Still the tale of the domestication of the horse is not
complete. Over a vast tract of northern Africa, including the
Sahara, there are thousands of rock-drawings depicting
horses; in some of them the horses are shown yoked in pairs,
at the gallop, and drawing a two-wheeled chariot. They have
been studied, of course, and principally by a French scholar,
Henri Lhote, who, in the Tassili group of these drawings,
has found affinities with Mycenaean art.[1] This has led
him to suggest that the so-called Sea Peoples, who attacked
Egypt in 1221 B.C. in alliance with the Libyans, brought
with them horses (i.e. horse-drawn chariots), and having
failed to conquer Egypt, moved westward into the Fezzan.
That is only one of the possible explanations of the arrival
of the domesticated horse in North Africa west of Egypt.
Another, based on the supposed need to explain the size
and weight of European horses by a separate domestication,
I have already mentioned. This calls for a domestication in
southern Spain which would have affected North Africa.
Even if it be so, it remains true that the first domestication
was Asian and dates from the third millennium B.C.

[1] Lhote (1953).

Finally, something must be said about the most remarkable and one of the most useful of man-made animals: the mule. Who first had the perverse notion of mating horse and ass? Someone may have thought it not a bad idea to breed an animal as big as a horse but with the patient endurance of the ass. I am not sure why Zeuner says that the ancient Jews supposed the Edomites and Hosites to have invented mule-breeding, on the authority of Genesis 36:24. What is said there is that Anah, son of Zibeon, found some mules in the desert while tending his father's asses. There is no likelihood of 'natural' mules being produced where wild populations of horses and asses, or horses and onagers, overlapped, if they ever did; the species do not mate together of their own free will. I have already said something about mules in the earliest Greek literature. There is a surviving fragment of Anacreon (c. 570–485 B.C.), in which the Mysians are named as the inventors of mule-breeding. This is confirmed by a very different source, for in Ezekiel 27:14 it is written that the staples from Togarmah [1] were horses, mares, and mules. The Israelites had to import their mules because the Law forbade them to make such sterile crosses.

All one can say is that the mule was in use before 800 B.C. and that it was probably bred first in Asia Minor.

[1] Like Mysa, in Cappadocia.

3

Poultry

When, late in the nineteenth century, Victor Heyn, a thorough and conscientious scholar, studied the history of the domestic fowl and concluded, among other things, that it was unknown to the ancient Egyptians until post-Alexandrian times he was mistaken, owing to the fact that he was writing a few decades before Howard Carter excavated the tomb of Tutankhamen. Each generation of scholars, scientists and historians draws conclusions from available data. This entails errors such as Heyn's; for the alternatives are, on the one hand, to draw no conclusions, a timid and unsatisfying cautiousness, or to draw conclusions which are provisional, a solution to the problem which was not in Heyn's temperament. In practice conclusions are revised continuously, as more facts are made available by research and excavation and, in the case of the living past (for the past lives on in the genes), by genetical analysis.

The domestic fowl reached Europe from the East in the eighth century B.C. Introduced first into Greece, where it did not become common for over a century, it was unknown to both Homer and Hesiod, both of whom flourished after 800 B.C. and before 700 B.C. The first surviving Greek reference to this bird is in the writings of the gnomic elegist Theognis, a Megarian born in the middle of the sixth century B.C. Later in that century the cock appears as a commonplace of man's economic life, its bellicose nature being used in lessons by moralists. The founder of Greek comedy, Epicharmus, who was born on Cos but lived most of his life in Syracuse, made use of the cock in such lessons. The cock also makes an appearance in Pindar (518–438 B.C.), who compares the shameful victories of civil war to the battles of barnyard fowls. And in the *Eumenides* of Aeschylus (525–456 B.C.) Athene warns the Athenians against waging civil war, which she stigmatizes as cock-fighting.

Two things are evident from these literary sources. First, the domestic fowl, unknown in Greece *c.* 800 B.C., became familiar there between 700 and 600 B.C., or a shade later, so that in the following century it was taken for granted as a part of the farmyard scene. Second, from its beginnings in Greece its importance as a fighting cock, a gamecock, was no less than its importance to the kitchen. I have read somewhere that Themistocles encouraged the troops before the battle of Salamis by reminding them that even gamecocks risk their lives, not for hearth and homes but for glory only. His ornithology was as bad as his morality; and considering that chickens are among the silliest of living creatures, I should not have thought his argument had much force. But it serves our purpose by proving that by the early fifth century B.C. the domestic fowl was familiar to all.

How did the domestic fowl reach our own parts of the world? Greek colonists carried it to Sicily and southern Italy where it first appeared *c.* 500 B.C.; a coin struck at Himera in 481 B.C. bears on one side the figure of a cock, on the other that of a hen,[1] probably because the bird was sacred to Asclepius (one recalls the dying words of Socrates), Himera being famous for its healing springs. Incidentally, bearing out the literary evidence, although the bird appears on other Greek and Italian coins, none of them is earlier than 550 B.C. From Greek Italy the domestic fowl spread up the peninsula to Latin Italy. It is at least possible that we have here the case of the wine-vine repeated, and that there was a separate introduction into Etruria, so that the birds were introduced into the Roman world from the north as well as from the south. I have not discovered just how early the Romans began to use

[1] Heyn (1888).

the new poultry for taking auspices; at least as early as the Sabine war. They did it by throwing food to the sacred chickens; if these ate greedily, so that the crumbs fell from their beaks, the omens for the coming undertaking were good; if not, bad. Some men of sense had no time for such nonsense. During the first Punic war a Roman admiral, P. Claudius Pulcher, ordered the sacred chickens to be thrown overboard when they would not touch the offered food. 'If they will not eat,' he said, 'let them drink.' The chickens were not the only ones who got wet that day: the Carthaginians sank Pulcher's fleet.[1]

By the first century B.C., when Varro wrote *De re rustica*, a work on rural economy, poultry-keeping had become a scientific craft. The Romans, like the Greeks, valued chickens at first primarily as gamecocks, then as egg layers, and lastly for the table. This was not true everywhere: in some cultures, eating the bird was tabu, or at all events forbidden by law.

By the same date chickens had crossed the Alps and were being kept by some Celtic tribes, for instance by the people of the splendid La Tène culture. Nor can they have been much later in reaching Celtic England, for Caesar says in *De Bello Gallico* that the Britons had fighting cocks, and archaeological finds of chicken bones on several sites confirm his statement.

So much for the dissemination of the domestic fowl through Europe. We must now return to Greece in about 800 B.C. and see where the bird came from. There are some clues in literature: Greek comic poets call it the Persian bird; in *The Birds* of Aristophanes the cock is Medos, the Mede, and it will be recalled that Peisthetairos wonders how, being a Mede, he got there without his camel. There is enough evidence of other kinds—monumental figures, inscriptions and so forth—to show that the spread of poultry-keeping coincided with the expansion of the Persian Empire over Asia Minor and the Greek islands. This, however, does not mean that the domestic fowl was an Iranian bird.

[1] The story is in Cicero. The battle was Drepana, 249 B.C.

There are four wild species of the genus *Gallus*, but the principal ancestor of our domestic chickens, although some other species have contributed genes, is the Red Jungle Fowl, *Gallus gallus*. The habitat of this beautiful bird in the wild is confined to a range, mostly in India, from Tonking in the east to Kashmir in the west. Far from extinct in the wild state, it is still abundant in Indian woods. In its wild state the Red Jungle Fowl hen lays about thirty eggs a year; but if you remove them as they are laid, she may manage as many as eighty. In domestication, with careful selection, generation after generation, of the best layers, the birds have been induced to lay upwards of 250 eggs a year, an extraordinary feat of artificial selection.

The distribution of the Red Jungle Fowl in the wild corresponds very nicely with the region of its first known appearance as a domesticated bird. A gamecock appears on a seal found at Mohenjo-daro, one of the metropolitan regions of the Indus Valley Civilization, and from the same area come clay figurines representing chickens, and chicken-bones larger than those of the wild Jungle Fowl, a fair proof of breeding in captivity. Similar evidence comes from other sites in the same culture, from Harappa and Chambu-daro. In short, the Indus Valley people had domesticated the Red Jungle Fowl not later than 2000 B.C.; and there is at least some evidence that its first function in the service of man was to amuse him as a fighting cock rather than to lay eggs for him as a hen.

From the Indus Valley the domesticated bird spread into western Asia by trade—there is a second-millennium B.C. word for it in Sumerian.[1] Whether the birds were eaten in their earliest phase as domestic birds, I do not know. But following the Aryan invasion of the land of their origin, they became sacred creatures, and by 1000 B.C. it was forbidden to eat them.

When Tutankhamen's tomb was excavated it was shown

[1] Zeuner (1963).

that the domestic fowl had reached Egypt by 1350 B.C. at latest, by the finding of a picture of a cock. Until that find was made, a reference in the annals of Thothmes III to birds which lay eggs daily had not been taken very seriously; now, it is accepted as fair evidence that the Egyptians did, in fact, probably have domestic hens by 1450 B.C. But following this early introduction—and hence the mistake made by Heyn— the birds disappeared from Egyptian farmyards and mains. Zeuner suggests that because of its connection with the religious revolution under Akhnaton and Tutankhamen, chickens may have been anathematized under the restored old religion, and so exterminated. Reintroduction did not take place until very much later, in Ptolemaic times.

Once domestic fowls had reached Iran from India they quite soon became of religious importance there, as well as economically valuable. In the religion of Zoroaster: 'The cock is especially dedicated to Craosha, the heavenly watchman who, awakened by fire, awakens the cock in his turn. He, by his crowing, drives away the daevas, evil spirits of darkness, particularly the yellow, long-fingered Bushyaçta, the demon of sleep.' [1]

The cock is the enemy of devils and magicians; he helps the dog to guard the house and other animals; his voice scatters evil; and Ormuzd himself had commended the cock to the prophet Zoroaster. It is no wonder that the Persians took the cock with them wherever they went.

One can sum up: *Gallus gallus* was domesticated in the Indus Valley before 2000 B.C.; the domestic fowl thus created had reached the Iranian lands and Egypt by about 1500 B.C. During the next half-millennium it colonized the farmyards of Asia Minor and China. It was introduced into Greece during the eighth century B.C., into Sicily a century or more later, into Italy in the sixth, had reached all France by the third and landed in England *c.* 100 B.C., by which time it was all over Germany and into Russia. For sixteen centuries its

[1] Heyn (1888).

advance was halted by the known limits of the world, but in
the sixteenth century A.D. Chanticleer and his hen began
their conquest of America.

GEESE AND DUCKS

The association between man and goose is probably a good
deal older than that between man and jungle fowl. The
linguistic clue is a clear pointer: the word for gander/goose
is the same, with merely local variations, throughout the
whole Indo-European group of languages, from the Old
Irish *geidh* in the extreme west (cf. ge-ith, geese), *anser* (cf.
(g)anser, gander) in the Latin west, to Sanskrit *hansa* in the
east. The same is true for duck. The conclusion is that all the
Aryan tribes, as they broke into Europe and into India,
already had the goose domesticated,[1] and probably duck as
well, but that is not so sure. On the other hand, there is no
question in this case of a single early Central Asian domestication, followed by dissemination; for we are fairly sure that,
for example, both the Egyptians and the Greeks accomplished their own domestication of the grey-lag goose.
Furthermore, on the evidence of some pottery models of
ducks and geese, the people of the Neolithic Lung Chan
Culture in Hupei, China, had domesticated both birds before
the appearance of bronze in that quarter.[2]

The goose is not mentioned in the *Iliad*; but in the *Odyssey*
there is a small flock of geese belonging to Penelope, and by
the sixth century we come to Aesop with his goose which
laid golden eggs. Elsewhere, representations more or less
identifiable, and other archaeological evidence, suggest other
local domestications. The ancestors of the Aryan Greeks
may have had a domesticated goose with them when they
arrived; lost it, and domesticated another later. The whole
picture is extremely confused.

[1] Heyn (1888).
[2] Watson, in Ucko & Dimbleby (1969).

Why, then, this precocious universality? The answer probably is that catching and taming geese and duck is very easy. Whole flocks of grown birds can be netted and penned; and that, in the southern lands of the domestic goose, is what must have happened, since the wild goose does not breed in the south, the places of its winter migration, but in its true home in the north. But in the north there would have been an important difference, and one which accounts for the evidence that in this case the peoples of the north were, despite their relative backwardness, the forerunners. Wherever men formed settlements, or even temporary camps of more than three months' duration—time for geese and duck to lay, hatch the eggs and rear the chicks—by rivers, estuaries, lakes, it would be easy for the boys of the community to find the nests of wild geese and duck, rob them of eggs or, better, of unfledged chicks whose wings could be clipped. That, or something very like it, must have been the beginning of domestication. And if it be a true hindsight, then one can say that it is an argument for an early north-Eurasian and equally early east-Asian domestication, before the Mediterranean ones, since the north was the only part where geese and duck could be found nesting.

Zeuner (1963) has a passage which, I think, bears on this:

The Nile Delta abounds in water-fowl and among them are about half a dozen species of geese. Of these, the grey-lag goose and the white-fronted goose were kept in captivity. It is difficult to decide whether these species were fully domesticated or kept captive, at any rate in Old Kingdom times. A fifth Dynasty relief in the Berlin Museum shows the stuffing of geese, Hyaena-fashion.[1] This proves their exploitation. But it must have been easy to catch masses of these birds with

[1] Hyaena-fashion: the Fifth-Dynasty Egyptians and later Palestinians and Syrians reared young striped hyaenas and fattened them for table by immobilizing them by tying the feet and force-feeding them with meat, mostly birds. A better analogue in this case would be 'Strasbourg-goose fashion'.

nets, and such scenes are shown frequently. In addition, eggs were collected. By New Kingdom times, however, the grey goose had become completely domesticated. It should be noted that the north of Europe had the domesticated goose at the same time. It is conceivable, therefore, that the complete domestication of the grey-lag goose in Egypt was due to influences from the north; on the other hand the species may have been actually breeding in Egypt in antiquity.[1]

Well, however they got them, the Egyptians certainly had domesticated geese before 1400 B.C., and so, at the same time or earlier, did the peoples of Mesopotamia.

As for duck, there is some evidence that Mesopotamia was a centre of domestication; but it is questionable evidence, for it turns on whether figurines of duck-like birds found by excavation are really duck or whether they are perhaps geese.

The wild species involved is *Anas platyrhyncha*, mallard; and birds of this kind seem to be figured in the so-called duck-weights—weights carved in hematite to represent duck, if they are duck, asleep with the head under the wing. As this mallard was certainly not domesticated in Egypt; as domesticated duck were introduced into ancient Greece and remained scarce; and as the Lung Chan duck of Hupei were too remote to have reached the Mediterranean west early enough to account for their earliest appearance as domesticated birds in that region, it does indeed seem that domestication must have been Mesopotamian. But there is another and very interesting possibility: Hallstatt-La Tène excavation sites at Sigmaringen, Germany, have yielded fibulae decorated with duck-heads and duck. Now this is not cogent evidence that the Bronze Age Germans had domesticated duck, but in view of what has been said above about the ease of taking unfledged birds for taming in those parts of the world where duck nest and breed, it is at least a possibility.

[1] If 'eggs were collected' does it not follow that the geese must have been breeding there?

PEACOCKS

Peacocks were domesticated for their looks, a rare occurrence. It is true that peacocks were eaten, and were, in a few places, farmyard birds. I myself have eaten peacock and very nasty it is, excessively tough and with no distinguishable flavour. (The same is true of swans.) The only people who have made a point of eating peacocks have been those concerned with conspicuous consumption—some of the Hellenistic *nouveaux riches* and, in the Middle Ages, the gaudier kinds of princes and prelates.

There are two wild species: *Pavo cristatus* of India, and *P. muticus* of Burma and Indonesia. Indian peacocks must have been domesticated quite early, for they formed an element of tribute or gifts sent from Indian princes to Tiglath-Pileser VI, whereby they were introduced into Mesopotamia. The people of southern Arabia, who traded across the sea with the Dravidians who were the first to learn how to make maritime use of the monsoon winds, had peacocks from India and used them in paying tribute or making gifts to the Babylonian princes. Meanwhile, peacocks had spread into Media, and it was from the Medes that the Greeks had them. On the island of Samos they became sacred to Hera. From Greece this fantastic-looking bird spread throughout the Mediterranean basin and up into inland Europe, there to become the *pièce de résistance* of the feasts given by millionaires, and an ornament of great, formal gardens. It was not displaced as a status-food until the introduction of the turkey.

THE TURKEY

The turkey reached England, let alone Spain, before Cortés had even finished the conquest of the Mexican Aztec Empire, in 1524. Did Kipling write, or was he quoting, the verse which runs:

Turkeys, Heresy, Hops and Beer
Came into England all in one year?

Columbus and his men must have been the first Old World people to see turkeys; for there were plenty on the islands, domesticated birds, since the wild habitat was the mainland of Mexico and the southern part of what is now the United States. Cortés and his people were, however, the first to notice these great birds, and could hardly fail to do so since they report flocks counted by millions. They took the turkeys for some kind of peacock, or at all events called them *gallopavo*, peacock, in their accounts of them.

That this bird has long been domesticated and was bred in enormous numbers is quite clear from the accounts of the early historians of the Conquest. Both the Aztec writer Ixtlilochitl [1] and the Spaniard Father Torquemada [2] vouch for this. Torquemada came into possession of the account books of Moctezuma's palace and found that the household consumed 8,000 turkeys at one marketing. Of course, that household was a military establishment and there were many mouths to feed. But the Emperor's housekeeper was not being extravagant; turkeys were cheap. The enormous number of turkeys bred is easily accounted for: apart from game, the Aztecs had only two sources of meat under their control, turkey flocks and domestic, edible dogs.

There can be no doubt, then, that the turkey was domesticated in Central America, and surely long before the rise of the Aztecs; but archaeological finds have not yet been such as to enable us to say when that domestication occurred or by which of the great Central American peoples it was accomplished. The Maya came from too far south. Olmecs, Toltecs? One just does not know.

[1] Ixtlilochitl MS. quoted by Prescott (1843).
[2] Torquemada (1773).

PIGEONS

Although the pigeon—*Columba livia* is the only important species in the context of domestication—was used as food by man at least as early as 4500 B.C. (Halafian period in Mesopotamia), and domesticated in the sense of being kept in dovecotes before the rise of the Mycenaean Culture, it is chiefly interesting for a unique function in the service of man: as a swift carrier of messages.

To deal first with its rearing as a source of meat. The origin of the dovecote (the pigeon-tower, a stone column with nesting places for pigeons) is oriental. I cannot find an account of its earliest occurrence, but it is certainly either Asian or, just possibly, Etruscan; and, as certainly, prehistoric. So a great many peoples had been keeping pigeons for food before the Romans became the pre-eminent pigeon-fanciers. The Romans virtually invented the basest practices of what we now call factory-farming. For although the Fifth-Dynasty Egyptians immobilized and stuffed geese and hyaenas to fatten them for the table, if not to produce the fatty degeneration of the liver called *foie gras*, they did not, as far as I know, actually mutilate the creatures. But the Romans broke the wings and legs of their doves to immobilize them, and then stuffed them with chewed bread and other grossly fattening foods.

That this ugly practice was not generally continued in our own culture is believed by many to be due to the Jews, who valued the dove for sacrificial purposes, requiring none but flawless birds, and to the Christians who took over this Jewish respect for the dove and promoted it to be a symbol of heavenly love. That did not prevent the monasteries from breeding doves for table in enormous dovecotes, but at least the practices of mutilation were dropped. There is, however, a distinct possibility that we do not after all owe dove-worship to the Judaeo-Christian religious complex: the Jews and Christians may simply have taken over the dove, and the thoroughly tiresome tolerance of great flocks of this bird in temple or cathedral precincts, from the worshippers of the goddess Astarte-Aphrodite. On the subject of the temple of Aphrodite-Ouranus, that is Astarte, in Syrian Ascalon:

There I found an innumerable quantity of pigeons in the streets and in every house, and when I enquired the reason, I was told that an ancient religious commandment forbade men to catch pigeons or to use them for any profane purpose. Hence the bird has become so tame that it not only lives under the roof, but is the table companion of man and is very bold and impudent.[1]

One of the oddest transferences in religious history is surely that of the dove from so lewd a goddess as Astorath-Astarte-Aphrodite to the Holy Ghost of the Christians; but it is a fact that the pigeon did become symbolic of the Third Person of the Trinity.

Now, as to the use of pigeons as mail-carriers. Just how early their homing instinct and their magnificent endurance and tremendous turn of speed in flight—an average of 60 m.p.h. over great distances is not unusual—were discovered seems to be unknown. But the first recorded instance of the use of pigeon-post was by Rameses III in Egypt, in the year 1204 B.C. When he ascended the throne he sent pigeons north, south, east and west to announce the fact. Zeuner (1963) thinks it unlikely that the true carrier-pigeon existed so early, but I do not know why; once the homing instinct had been noticed—as it must have been, or Rameses could not have arranged for his doves of annunciation to fly to the four cardinal points—their use as messengers would have been obvious. I think that even in Genesis 7:9 and 6:12 there is a clear hint that its twelfth-century B.C. author distinguished the dove for its homing instinct; for whereas the raven which Noah released on the fortieth day of the Flood

---

[1] Philo Judaeus [30 B.C.–A.D. 45].

did not return but flew round in circles until the waters ebbed, the dove released seven days later, and again seven days after that, returned on both occasions.

By the fifth century B.C. this very ancient discovery of how to make use of a bird's homing instinct had reached Athens, for the Athenians and Thracians were using carrier-pigeons in their commercial and political relations. Again, when Taurosthenes of Aegina won a victory at Olympia he sent the good news to his father by carrier-pigeon which arrived on the same day. The Romans took this notion from the Greeks, and the Emperor Nero used pigeon-post to carry the results of the day's sporting events in the circus to his friends in the country. It is curious that this postal system appears not to have been used in war by any of the ancients. The Arabs were the first in that field, and under the Mamelukes in Egypt, and the Fatimid Caliphs, the breeding of racing-pigeons for mail-carrying was carried to a high level of perfection, with the proper use of pedigree recorded in stud-books.

GUINEA-FOWL

The guinea-fowl, *Numidia meleagris*, was the last of the familiar (nowadays less familiar) farmyard birds to be domesticated. In its association with man this bird has had a curious career. An African member of the *Gallinaceae*, it was discovered in the wild by the Portuguese when they began to explore parts of Africa in the late fifteenth century. They introduced it from West Africa to Europe—hence the English name 'Guinea'-fowl—and in that sense were responsible, from our point of view, for its domestication, all our guinea-fowl being descended from those Portuguese introductions. What no fifteenth-century European, except a few scholars, remembered was that the bird had once been domesticated by the ancients but had then been so completely lost as to be unknown in the Middle Ages.

Given the name *melanargis* (corrupted to *meleagris* [1]),
which simply means black-and-white, it reached Greece in
the fifth century B.C. If asked to guess where from, I should
say Carthage, or one of the Greek colonies in North Africa
which had had the bird from Carthaginian neighbours; for
the Carthaginians were the only people who had frequent
contact with those parts of West Africa from which the birds
came. The Greeks themselves seem to have had no idea of
the guinea-fowl's place of origin, and they attributed to it
fantastic qualities, such as weeping tears of electrum.[2]
Heyn (1888) quotes a writer called Mnaseas who knew where
the birds came from—the West African land of Sicyon; and
as that place was thought to be the main source of electrum,
those electrum tears in Sophocles are explained. The Greek
explorer Scyrax (fifth century B.C., but the *Periplus* attributed
to him was written about one hundred years later) identifies
the place and its guinea-fowl:

If one sails through the pillars of Heracles, keeping Africa
on one's left hand, there opens, as far as the Cape of Hermes,
a wide gulf called Kōtēs; in the middle of this gulf lies the
town of Pontion and a large, reedy lake called Kephasias;
there live the birds called *meleagrides*, and nowhere else,
except where they have been taken from that place.

I have no idea where Cape Hermes was, or the other places
mentioned. The Portuguese found the guinea-fowl in that
region of Gambia; it seems rather unlikely that Scyrax got
so far, although the Carthaginians may have done so. The
range of the bird may have been farther north. Nearly two
thousand years separate the two discoveries, and the Rio de
Oro cannot always have been as arid as it is now, for it is
rich in Neolithic remains.

At all events, I suggest that Carthaginian seamen first
brought home the guinea-fowl from West Africa, and that
it was domesticated in Carthage and from there distributed
to the Hellenic world.

[1] A more plausible explanation than Heyn's.   [2] Sophocles.

4

Cats and Polecats

The cats which live with us now are the most pampered
of all domestic animals; they are not, like dogs, under
discipline, and since they are not so clever as dogs, less is
expected of them. They are much more nearly as free as
wild animals than dogs are; they do not suffer the emotional
blackmail to which dogs are submitted; nor are they required
to do any work, although we are grateful to them if they
catch a few mice. Their principal function is to be decorative, and for them that is a sinecure.

It was not always so. One might almost say that the world
of men, with the honourable exception of the Egyptians,
owes cats a living as some compensation for the dreadful
things done to them in the past. Perhaps no animal ever
suffered such abominable cruelty at man's hands as did the
domestic cat in mediaeval Europe, and the poor brutes were
not much better off in China and Japan.

By some means not at all easy to follow, the cat, which
unfortunately for its kind reached Europe at the same time
as Christianity (a religion which has a very bad record where
animals are concerned), became identified with Satan. Now
Satan being the enemy of man, it became usual to mortify
him by maltreating cats, especially black cats. Throughout a
large part of Central Europe, Germany and Flanders it was
customary during Lent to kill and bury as many cats as
possible, or to bury them alive. During the festival of Easter,
in the Vosges mountains and in Alsace generally, cats were
regularly burnt alive. (One must, of course, keep a sense of
proportion: the Church also delighted in burning people

alive.) In the mountains of the Ardennes cats were thrown onto bonfires—alive, of course; or they were tied to the ends of long poles and roasted alive. The Celtic British refined on this by impaling cats on spits and roasting them over a hot fire.

All these and suchlike practices were, of course, magical rites: the intention was to drive out the Devil. Even at the time they were deplored by the more enlightened; the south Welsh legal code of Howel Dda, for example, included a law for the protection of cats from such hideous abuse.[1] But elsewhere in Europe, with a Church as terrified of resurgence of the Old Religion as is a modern United States administration of domestic Communism, only a very brave man dared risk suspicion of heresy by trying to save a cat from torture.

Even later than that most base period of abject superstition cats had a shocking time of it. In the sixteenth and seventeenth centuries, all over Europe and in America, there was a practice of immuring cats alive, to die, desiccate and remain ever on guard as a warning to the Devil—or perhaps, to local rats; or they were killed, mummified by drying, and put into hollow walls or under floor-boards, sometimes with a dead and desiccated rat in their mouths, apparently as a rat-scare.[2]

The rat-and-mouse association of cats is, of course, important; but the job of keeping down these vermin was only taken over from a more anciently domesticated mouser, at least in the Hellenic world. Most authorities have thought that weasels or ferrets—domestic names for polecats—were domesticated to deal with rabbits; but I think that literature shows a more ancient use. For example, at the beginning of the *Batrachyomachia*,[3] the Mouse tells the Frog: 'Two beings

---

[1] Spence (1945).

[2] Howard (1951).

[3] *Batrachyomachia*, 'The Battle of the Frogs and Mice', Greek mock-heroic epyllion *c.* fifth century B.C. Some scholars have tried to attribute it to Homer.

I fear above all things on earth, the Hawk and the Weasel, which have done my kind much harm; yes, and that painful, fatal, deceitful Trap; but most of all the Weasel which is strongest and even comes creeping into my holes.' I think that the association with the trap makes that a domestic weasel. And, note, no mention of cats.

Heyn (1888) made the interesting suggestion that the rather sudden spread of cats throughout the Hellenistic world in the fourth century A.D., corresponding as it does with the coming into use of the word *catus* [1] (thereafter universal in all European languages), may perhaps be connected with the invasion of Europe by '... a hitherto unknown voracious rodent coming from Asia, the Rat, *Mus rattus* ...', which, by his account, came into Europe with the Teutonic barbarians having, like them, been disturbed by the enormous movements of such people as the Huns in Central Asia.

Then why, if the cat came as a boon to help us defend our larders and barns against rats, did we soon begin to maltreat our ally so hideously? The answer lies in its occult associations, the origin of which would seem to be the cat's status as a god in Egypt. True, the dog also was a god there; but man had already had the dog for thousands of years, whereas the cat was new and arrived with an ambiguous reputation. That reputation was strengthened by the equivocal character of the animal itself: nocturnalism was always suspect; so was association with the moon, an association which was very bad for the cat outside Egypt, right across the world from western Europe to eastern China and Japan.[2] Then, deeply disturbing, there are the electric charge in the cat's fur, producing blue sparks in the dark; and the water-clear yet enigmatic eyes. Both phenomena bore out the beast's guilt-by-association with the avatar of an anti-Christian god, that is to say, a

[1] *Catus* is specific; former words, e.g. *feles*, *meles*, are not, and seem usually to mean polecats. (*Opus agriculturae*, fourth century B.C.).
[2] *See*, e.g., Vulson de la Colombière (1644).

devil. Even as late as the seventeenth century a sensible man could write: 'The cat is more harmful than useful, its caresses more to be dreaded than desired, and its bite fatal.' [1] No doubt, too, the strange phenomenon of cat-allergy, causing asthmatic crises and even fainting, appeared as early as the cat in our midst. Finally, there is the animal's self-sufficiency: men, used to being worshipped and slobbered over by dogs, could not be expected to like an animal of whom a great French naturalist once said, referring to the cat's apparently affectionate ways: 'Votre chat ne vous caresse pas; il se caresse à vous.'

The cat's role has always been equivocal: devil or god? In the illuminated MS. known as Queen Mary's Psalter, the serpent-woman Lilith, responsible for our downfall by the tempting of Adam, is shown with the head of a beautiful woman but the body of a cat.[2]

On the other hand, Mills (1820) has this extraordinary story:

At Aix in Provence on the festival of Corpus Christi the finest Tom-cat in the country, wrapped like a child in swaddling clothes, was publicly exhibited in a magnificent shrine. Every knee was bent, every hand strewed flowers or poured incense, and in short the cat on this occasion was treated like the god of the day.

In both China and Japan the connection with the occult is just as marked and has given rise to a whole literature of fantastic tales, yet not, I daresay, as fantastic as the following:

For twenty-five years an oral addition to standing orders of the native guard at Government House near Poona had been communicated regularly from one guard to another on relief, to the effect that any cat passing out of the front door was to be regarded as His Excellency the Governor and to be saluted accordingly. The meaning of this was that Sir Robert Grant, Governor of Bombay, who died there in 1838 and on the

[1] Palliot (1664).
[2] Conway (1879).

evening of the day of his death a cat was seen to leave the house by the front door and walk up and down a particular path, as it had been the Governor's habit to do after sunset. A Hindu sentry had observed this and mentioned it to others of his faith, who made it a subject of superstitious conjecture, the result being that one of the priestly caste explained the mystery of the dogma of the transmigration of the soul from one body to another, and interpreted the circumstance to mean that the spirit of the deceased Governor had entered into one of the house-pets.

It was difficult to fix on a particular one, and it was therefore decided that every cat passing out from the main entrance after dark was to be treated with due respect and proper honours . . .[1]

But I suppose all that has not much to do with the cat's origin in our homes.

There were few or no cats in Europe at large until Roman Imperial times; and the amount of archaeological and literary evidence shows conclusively that it was the Romans who took the domestic cat into all the lands of their Empire north of the Mediterranean. Some have argued that the Romans had the cat early from their more civilized Etruscan neighbours to the north; but the evidence on which they relied, as to the presence of the domestic cat in Etruscan cities, has been so effectively demolished by Zeuner (1963) that it is now quite clear that the Etruscans had no cats. Yet doubtless there were cats in Rome before their introduction to such outposts of the Empire as Britain. The Romans could have had them from Greece by way of Greek Italy, or directly from Egypt.

The Greeks had a few cats, as rare curiosities, in the fourth century; probably they began to appear there as early as 500 B.C. Zeuner describes and figures a marble relief from Poulopoulos (c. 480 B.C.) showing men confronting a leashed dog with a leashed cat, and obviously waiting with keen interest to see what will happen. But introduction to Greece cannot have been much earlier, for the cat was not a common-

[1] Gordon (1906) quoted by Van Vechten (1921).

place until very much later; Aristophanes (257–180 B.C.), for example, still has the weasel, not the cat, as domestic mouser. The cat was earlier in Crete, possibly as early as 1100 B.C. When one thinks of how quickly cats multiply with us, to the point where their number becomes an embarrassment, they seem to have taken a remarkably long time to establish themselves firmly in a new country. Thus, an ivory statuette of a cat found at Lakish in Palestine is regarded as evidence for the presence of the domestic cat in that country as early as 1700 B.C.; yet there is no mention of cats in the Bible, the earliest books of which were written 500 years later. Still, there is a cat in the Apocrypha. Could it have been deliberately excluded from the sacred writings as unclean? Hardly, for as unclean it ought surely to have been listed in Leviticus with other unclean creatures.

The only place in the world where domestic cats were abundant, a part of everyday life, before the above date, was Egypt; and there it was mouser, bird-hunter, pet and god.

The number of domestic animals in Egypt is very great, and would be still greater were it not for what befalls the cats. As the females, when they have kittened, no longer seek the company of the males, these last, to obtain once more their companionship, practice a curious artifice. They seize the kittens, carry them off, and kill them, but do not eat them afterwards. Upon this the females, being deprived of their young, and longing to supply their place, seek the males once more, since they are particularly fond of their offspring. On every occasion of a fire in Egypt the strangest prodigy occurs with the cats. The inhabitants allow the fire to rage as it pleases, while they stand about at intervals and watch these animals, which, slipping by the men or else leaping over them, rush headlong into the flames. When this happens the Egyptians are in deep affliction. If a cat dies in the house by natural death, all the inmates of the house shave their eyebrows. . . . The cats on their decease are taken to the city of Bubastis, where they are embalmed, after which they are buried in certain sacred repositories. . . .[1]

[1] Herodotus, ii. 66.

The punishment for killing a cat deliberately was death; and there are a number of well-authenticated stories of foreigners who, having accidentally killed a cat, were lynched by infuriated mobs, despite the protection of the authorities.

Yet in Egypt the cat was not nearly so ancient as a domestic animal as were most other animals possessed by the Egyptians. It was domesticated in historic times; in fact there is no clear and unambiguous evidence for domestication before the Fifth Dynasty, and only thereafter did cats become commonplace. Neither Old nor Middle Kingdom sites have yielded evidence for a domestic cat. But with the New Kingdom in the sixteenth century B.C. it is very much in evidence; cats are even shown helping to hunt birds; and by that time also they had become sacred to the goddess Bubastis, or Pasht.

That goddess is invoked every time we call our own house cat: *puss*, *pussy* are corruptions of *Pasht*. For the word *tabby* two possible explanations are offered. The most usual derives from *utabi*, a Turkish word for cat; it is supposed to have come with a new and strikingly good tabby race from Turkey. The other explanation is that the word derives from the name of a striped fabric first manufactured in the twelfth century and named after an Ommayad prince called Attab.[1] This sounds far-fetched and seems too late. *Catus* or *cattus* is from the Greek *kattos*, which the Greeks must have had, with the animal itself, from Egypt. It has affiliations with the Arabic *guttah* (Italian *gatto*) which itself comes from a much older language, Berber.[2]

So it would seem that the cat was domesticated in the second millennium B.C., in Egypt. Why, then, did such an enormous period of time elapse before its introduction into other cultures? The answer appears to be that the Egyptians, not liking the casual way in which this sacred animal was treated outside Egypt, forbade and prevented the export of cats; there are even tales of travelling Egyptians who,

[1] Hitti (1956).
[2] Keller (1909).

coming across one of the rare cats outside Egypt, bought it and took it home, where it would be properly treated.

The modern cat is not derived solely from this Egyptian domestic cat. There are probably at least three species involved: 1. the North African (including Egyptian) *Felis lybica*, the wild cat which, by making its way boldly into Egyptian houses in search of mice, found itself welcomed, settled down with man and was rewarded with deification; 2. the Manul—*Felis manul*—the wild cat of the Asian steppe; 3. the European wild-cat, *Felis silvestris*, a woodland, climbing animal that still survives in Scotland. (*Felis catus* is properly applied only to the domestic animal.) As soon as the Egyptian cat was well established in a new place, some of the domestic population went feral, so that cross-breeding occurred with the local species. This had two effects: it not only brought new attributes into the lines of domestic cats, but also introduced genes of *F. lybica* into the wild cat populations outside North Africa. Special markings or absence of markings are the result of selective breeding. The wild cats are blotched tabbies; the domestic tabby only is banded rather than blotched. The Abyssinian cat is not, according to Zeuner, Abyssinian, but may owe its appearance to a wild race of Sardinian *F. silvestris* in which the markings tend to disappear. The long hair of Persians and Angoras is the product of selective breeding; it has nothing to do with geographical races of wild cats, and probably nothing to do with Persia or Angora. The Siamese cat, first seen in England in 1884, may possibly be Siamese, but may equally well be Indian. The Egyptian cat arrived in India by way of Babylonia early in the second century B.C., and became crossed with a local race of *F. lybica* called *ornata*; the characteristic colour scheme, beginning with white in the kittens, is known in other small mammals in India, for example in a race of rabbits. On the other hand, the kinked tail is characteristic of many cats in Burma, Thailand and Malaysia. As for the Manx cats, taillessness is the result of a mutation which is not confined to the Isle of Man but occurs everywhere and is common in the Far East.[1]

The spread of the domestic Egyptian cat led to decline in the fortunes of the domestic weasel or ferret—the tamed version of the wild polecat. I have already quoted from the *Batrachyomachia* to show that the house-predator of which the Mouse was afraid in the sixth century B.C. was the weasel, not the cat. The same was true for centuries afterwards, for, as I have said, the cat did not increase quickly on first intro-duction. But as it did increase, it must have forced out the poor weasel in more senses than one; it probably ate the house-weasel. But proof is not wanting that the house-weasel managed to last for a long time after the introduction of the cat. Thus, in *The Wasps* of Aristophanes a character asked to tell a domestic story replies: 'Anything to oblige. Let's see, once upon a time there was a mouse and a weasel . . .', just as we might say 'a mouse and a cat'. Still later, in a play by Plautus [2] (254–184 B.C.), a weasel, not a cat, catches a mouse at the feet of one of the actors.

Admittedly it is hard to be sure which animal the classical writers are talking about, because the names they use are far from specific. But Heyn insists on linguistic grounds, where despite other faults he was very sure-footed, that in the ances-tor of the Grimalkin story, Fable 32 of Babrius, the 'cat-woman' is not a cat-woman but a weasel-woman. Babrius was versifying Aesop in the second century A.D., so that he is probably less reliable for his own time than for the fifth century B.C., if he was sticking close to his original, a point which Heyn seems to have missed The same philologist will not allow the Greek *ailouros* or the Latin *felis* to mean cat, unless by a late transference. The words mean some kind of house-marten or house-weasel; only *katos, catus*, really does mean a cat, and they hardly appear until Palladius (*De re rustica*), that is, until the fourth century A.D.

[1] Zeuner (1963).
[2] Quoted by Heyn (1888).

1

2

3

1. Wolf (*Canis lupus*), ancestor of our dogs, first associated with man in a hunting partnership. Still found wild in small numbers in parts of Europe and North America.

2. Siberian Wild Dog. The wolf kinship is apparent; at this stage it is difficult to distinguish wild from feral animals.

3. Maned Wolf (*Chrysocyon jubatus*): peoples of South America had their own dogs before the coming of the Europeans. This animal, still wild in Brazil and Argentina, is a probable ancestor.

4. Singing Dog, New Guinea, another wild dog whose wolf kinship is manifest. A possible beginning of domestication is the adoption of orphaned wolf puppies by the children of Palaeolithic communities.

4

5

6

5 and 6. Coyotes (*left*) and jackals readily enter villages to scavenge for food thrown out by the villagers. This useful service led to association of man and jackal, which thus became an ancestor of the domestic dog. As in the case of the wolf, domestication begins with a free and mutually valuable partnership.

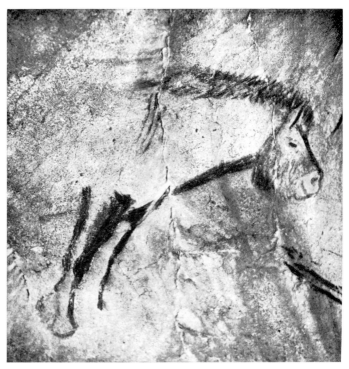

7

7. Palaeolithic cave-drawing of a horse, Niaux, France: note the points of resemblance with . . .

8. . . . a wild horse about twenty thousand years later, from a species (Przewalski's wild horse) ancestral to the domesticated horses.

9. Although often reputed untameable, the Onager was domesticated in Central Asia before the horse and was one of the ancestors of the asses of draught and burden used by the proto-civilizations of West Asia.

8

9

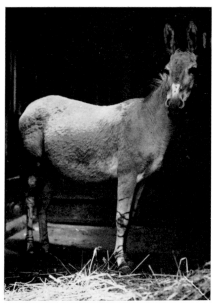

10

10. Wild Ass of Somalia, ancestral to the domesticated asses of North Africa which are older in man's service than the true horses.

11. Kiang, or Asiatic (Mongolian) wild ass. Possibly crossed in prehistoric times with the Somali or Nubian ass.

11

Zeuner seems to take Leviticus 11 : 29–30 as evidence for domestication of the polecat as early as 1000 B.C.; I do not agree, for these verses simply list unclean animals with no implication that they are tame; as far as I know, nobody ever domesticated lizards, which appear in the same short list. In any case, what was this animal? The New English Bible calls it a mole-rat,[1] and adds a footnote saying 'or weasel'. At all events, the polecat was certainly domesticated before 500 B.C.; and although its job of keeping the house clear of mice was taken away from it much later by the cat, it retained its importance, for it was the best known rabbiter.

The earliest account of its use against rabbits is in Strabo (64 B.C.–after A.D. 21). He tells of Libyan ferrets bred for the work; of their use against a plague of rabbits in the Balearic Isles; and of how they were muzzled before being slipped into the warrens. This use of ferrets is the only one which had kept them going in domestication, although they are excellent mousers.

A related, but much bigger, animal was domesticated— perhaps just short of being bred in captivity, but I am not convinced of that—by the ancient Egyptians: the mongoose, or ichneumon. How early this was accomplished does not appear; but it is surely true, as Zeuner says, that it began with the mongoose slipping into men's houses after the mice, and, above all, after the snakes. It was this skill in destroying snakes that led to domestication. I do not know why Zeuner says the use of the mongoose was confined to Egypt. I suppose he means that it did not become, like the cat, a universal animal. True, but surely Kipling's Rikki-tikki-tavi was Indian?

The domesticated mongoose reached Rome, but apparently only as a pet. Martial (first century A.D.) says that Roman ladies liked to have ichneumons as pets. In view of the constant battle which we have still to fight with the rats

[1] Defined by O.E.D. as 'any myomorphic rodent of the family *Spalacidae*'.

which came into Europe with the Germans, it is surprising
that we have not taken much greater advantage of this incomparable snake-killer, which is a much more efficient ratter
than cat or dog. Early in the 1930s a great publishing house
had one of its paper warehouses overrun by rats. Every
other means having failed, they tried a mongoose. It had
cleared that warehouse of rats within four months, after
which the firm had the job of finding food for the mongoose.

It was, again, the Egyptians, whose oneness with nature,
with animals, has never been equalled by any subsequently
civilized people, who accomplished the domestication of one
of the big cats—the Cheetah, *Acinonyx* (formerly *Felis*)
jubata. Whether it was fully domesticated in the sense of
being bred in captivity, I do not know; Zeuner thinks not
(but see below). At all events, it was tamed and trained not
later than the fifteenth century B.C.—the approximate date
of the tomb of Rekhmere which yielded a picture of a cheetah,
wearing collar and lead, being led like a dog.

Cheetahs were used for hunting gazelles. They are not
very difficult to tame. A cheetah with five enchanting cubs
in the Kenya Game Reserve allowed a friend and myself to
approach within a yard or two before she got up and, followed
by her litter, walked slowly and calmly away. There are, of
course, problems in keeping cheetahs in urban-industrial
contexts: a friend who offered to sell me his cheetah, cheap,
throwing in collar and lead, was forced under questioning to
admit that, out for a stroll by itself in one of the leafier
suburbs of Manchester, it had eaten a neighbour's Pekinese.
A pack of cheetahs in the cut'em-down-and-hang'em-up-to-
dry country would add colour to the English hunting scene—
but these are dreams. . . .

From the number of pictures of tame cheetahs found on
Eighteenth- and Nineteenth-Dynasty sites in Egypt, it is
clear that the hunting cheetah was by that time not much less
common than the hunting saluki.[1] They are much faster

[1] Keller (1909).

than greyhounds over a short distance, but have less staying power.

From Egypt domesticated cheetahs were sent, but comparatively late, to Mesopotamia. The Assyrians made frequent use of them for hunting, and since the cheetah is not native to that country, Egypt must have done a considerable export trade in tamed cheetahs. But the animal is native to India—or rather, was; it has now been exterminated there—so that once the Indians of the north, and their neighbours the Afghans and others, had conceived the idea of taming cheetahs for hunting, and perhaps even of rearing them in captivity,[1] they were able to draw upon the native cheetah population.

Despite Zeuner's great authority I find it hard to believe that the cheetah was never wholly domesticated, to the point of being bred and reared as a domestic animal, in view of two passages from Marco Polo's *Travels* (thirteenth century) on the subject of the Great Khan Kublai's hunting arrangements:

Frequently, when he rides about his enclosed forest, he has one or more small leopards carried on horseback behind their keepers; and when he pleases to give direction for their being slipped, they instantly seize a stag, or a goat or a fallow deer, which he gives to his hawks, and in this manner he amuses himself . . .

It is difficult to imagine a horse remaining manageable with a cheetah on its haunches, unless the animals had been used to each other all their lives. The second passage reads:

The Grand Khan has many leopards and lynxes kept for the purpose of chasing deer, and also many lions which are larger than the Babylonian lions, have good skins and handsome colour, being streaked lengthways, with white, black and red stripes. They are active in seizing boars, wild oxen and asses, bears, stags, roebucks and other beasts that are

[1] Zeuner (1963) says not. Cheetahs, though threatened with extinction, are still found wild in Iran where they are strictly protected.

the object of the sport. It is an admirable sight, when the lion is let loose in pursuit of the animal, to observe the savage eagerness and speed with which he overtakes it.

What those 'larger' animals were, heaven knows. They sound more like tigers than lions; and I am not suggesting that they had been domesticated, although I suppose they might have been kept to breed in the vast hunting parks which the Mongols created. But the 'small leopards' are, of course, cheetahs, and it seems to me at least possible, even probable, that they were quite as domesticated as hunting dogs.

5

Sorting the Sheep from the Goats

Moralists have always had a good deal of difficulty in sorting the sheep from the goats, a proverb, by the way, whose implications clearly reveal the distinction which primitive shepherds made between the merits of these two animals. But zoologists, with nothing to work on but fossil and sub-fossil bones or horn-cores from excavations, are in even worse trouble. Sheep and goat skeletons are very nearly indistinguishable from each other; and they may become quite indistinguishable when you have nothing but ancient fragments to examine. To decide, from such exiguous and, on the whole, poor bone material as archaeology makes available, whether the animal was a goat or a sheep, and whether or not it was domesticated, is so difficult that some authorities hold that conclusions drawn from such decisions are bound to be unreliable and therefore misleading.[1] It is even possible to include in the sheep/goat group the bones of other similarly sized *Bovidae*, some gazelles, for example, and antelopes. But this is perhaps an excessively gloomy view; scientific integrity demands extreme caution in drawing conclusions, but surely something can be made of the available evidence.

The possible ancestors of our domestic sheep are numerous. Some obvious ones are *Ovis musimon*, the European Moufflon, native still to Corsica and Sardinia; *Ovis orientalis*, the Asiatic Moufflon, native to south-west Asia and Cyprus; and the Urial, *Ovis vignei*, whose ancient range was from eastern Iran to Tibet, and north into Turkestan. Then there is a Siberian sheep called Argali, *Ovis ammon*; and wild

[1] Reed (1960).

sheep are not even wholly confined to the Old World, for there is a North American sheep called *Ovis canadensis*. Could this last be a Siberian sheep which crossed the ancient Behring isthmus? At all events, these are the species according to one respectable authority;[1] others distinguish many more species where he sees only geographical races.

Coming to the goats, the data are not so complicated; since there is no evidence that the ibex, *Capra ibex*, was domesticated except experimentally in Egypt, it may be that there is only one possible wild-goat ancestor, *Capra hircus aegagrus*, a native of south-west Asia. This animal is of particular interest here because, on the evidence at present available, it must have been the first of all the ruminants to be domesticated. The best way to look at it is by glancing first at the history of the goat in domestication.

Now this reveals that every Oriental and European people has had domesticated goats since the dawn of history, that is, since before the invention of even the most primitive form of writing. There is, thus, no question of being able to trace its domestication within historical times. Nor would there be even were it not established that goats, like dogs, were our associates in the making of historical civilization. For there is one remarkable fact which is almost conclusive: the goats of the Canary Islands are of south-west Asian domesticated stock, although they have become feral there. From all, even the earliest, accounts of the islands it is clear that the native Guanches had always had these goats; and the Guanches were still in the Late Neolithic phase of culture when Jean de Béthencourt began the conquest and settlement of the islands by Europeans in 1402. I say that this is 'almost' conclusive, because it is just possible that somebody took goats to the islands long after their settlement by man from north-west Africa; but it is at least more likely that the goats went there in the first place as domestic animals, with Neolithic man.

[1] Lydekker, quoted by Reed (1960).

Domesticated goats of the almost universal 'Bezoar' stock, of which there are several geographical races with differences in size, shape of horns, etc., were herded by Neolithic Swiss Lake Dwellers, and by people in the Neolithic stage in southern Germany. The Hungarians of the Hatvan and Toszeg Early Bronze Age cultures had herds of bezoar goats. Crete had herds of goats in her Neolithic phase, long before the appearance of sheep in Late Minoan (1600 B.C. onwards). Pre-Dynastic Egypt, and perhaps even pre-agricultural Egypt, had domesticated goats. We shall go into that in detail below; here it may be said that there were two distinct 'breeds' of domestic goat in Egypt well before 3000 B.C., and the same is true for prehistoric Palestine and possibly Greece. Bones of what may have been goats have been found associated with a Mesolithic culture (Tardenoisian) in Belgium, although the animal was more probably a wild ibex.[1]

That will be enough to show that domesticated goats of south-west Asian stock had been introduced over a vast range of territory in the New Stone Age, before the people of the territory in question had the use of bronze or even of copper. The question is, where did they come from? And, before we come to that, how do experts tell the remains of domesticated from those of wild goats?

The answer to this last question is the same as that given elsewhere (but in other contexts) in this book. Whether one is considering animals or plants, morphological changes occur in domestication, as also do mutations which are preserved by selection and segregation. In the case of the goat I cannot do better than quote a short passage from Dr Reed (1960) on the subject:

In goats such mutations seem to have affected the horns, and their bony cores, particularly of the males. In wild goats the horns are scimitar-shaped, as long as a metre in an old

[1] Zeuner (1963).

male, and embossed with bumps on the front. After death the keratin of the horn disintegrates unless it is kept absolutely dry, so that it is invariably only the bony core which is recovered archaeologically. The cores are some two-thirds as long as the external horn sheaths, have a sharp keel anteriorily, and are irregularly quadrangular in section. Changes towards an almond-shaped cross-section, or flattening of the medial surface, are presumptive evidence of domestication. Twisting of the cores indicates that the goats were screw-horned as most modern domestic goats are; this condition is unknown in wild *Capra hircus*. The earliest domestic goats in South-west Asia were all scimitar-horned, but this type was generally replaced by screw-horned goats by the Bronze Age.

Three archaeological sites compete for the honour of having yielded the earliest domestic goats.

First there is a claim for the bones and horn-cores recovered from what archaeologists call Natufian III-IV levels at al-Khaim in the Wadi Kharaitun, which runs from a point near Bethlehem down to the Dead Sea. But this claim is based not on the bones themselves, but on the striking fact that whereas there are no signs of goats at all in the earlier levels, they appear suddenly in the Natufian III-IV levels. In other words it does seem that they must have been suddenly introduced, and hence, it is argued, they must have been domestic. The bones themselves are those of the wild bezoar or an equivalent domestic animal (*Capra hircus aegagrus*). For it takes time for morphological changes to occur and be noticeable following domestication, and one would not expect to find them at the first stage of domestication; so that even if the goats in question were domestic, in this case they would be indistinguishable from wild goats.[1]

The date of these finds could conceivably be as early as 10,000 B.C. But the claim for domestication has been very strongly criticized; it is held that whatever the explanation for the sudden appearance of goats in this culture and place,

[1] For this discovery and the claims made for it see Vaufrey (1951).

it cannot be accepted as evidence for domestication, and nowhere else have goats been found associated with such early Natufian remains.[1]

Next we come to Zeuner's own find of goat-horn cores in pre-pottery Neolithic levels at Jericho. The radio-carbon date is nearer to 7000 than to 6000 B.C.; the cross-section of the cores is very like that of Bronze Age, scimitar-horned and unquestionably domesticated goats, and Zeuner believes that on the evidence available to date his Jericho animals were the first domesticated goats and the first of all ruminants to be domesticated anywhere.

Goat remains were recovered from a site at Jarmo in Iraq, and Reed thinks that they are remains of domesticated goats: here the horn-cores are not only flattened or, like the Jericho ones, almond shaped in section, but also show some signs of twisting.[2] Jarmo is also a pre-pottery Neolithic site, but later, say about 5000 B.C.

Another site which has yielded remains of domesticated goats is the Belt Cave near the southern shore of the Caspian Sea in Iran. The date for these is about 6000 B.C. (C[14] date 7790 ± 300 BP).[3] Dr Carleton Coon, who made this discovery, believes that even thus early goats were domesticated and were being milked, but it is not clear how milking can be deduced from bones.

Later than these dates, evidence for goat domestication becomes a little more plentiful. One very interesting point is that where clearly identifiable goat and sheep remains are found on the same site, the goat remains are on the lower, that is the earlier, levels. This is not invariably the case; but where the order is reversed the dates seem to me late enough for both animals to have been domesticated for some time elsewhere, and therefore to have been introduced, in which case the order is not significant.

[1] Zeuner (1963) and Reed (1960).
[2] Reed (1960).
[3] Coon (1951).

In Egypt, early though they be, goat remains are later than in Palestine (Jordan), Iraq and Iran, the earliest being around 4000 B.C. But then it is not only possible but even probable that there is, as it were, a fortuitous gap in the records: the domesticated goat can hardly have taken 2,000 years to make its way from Palestine into Egypt. But from our point of view it does not matter: there is not much room for doubt that the goat was domesticated in south-western Asia, in Palestine and Iran, not later than 6000 B.C. and perhaps as early as 7000 B.C.

It seems to me that there can be little doubt that this early domestication of the goat has had some pernicious consequences for the ecology and perhaps even the climate of those regions. Goats are browsers; they are tree-destroyers, and the damage they have done in the Near and Middle East during some eight thousand years has been considerable. What a pity that Neolithic man did not, rather, make a beginning with the sheep, which has replaced the goat in most economies.

There is at present no question of being able exactly to identify the ancestral wild species of sheep; several are probably involved and, in the simplest terms, sheep populations are now too mixed to be sorted out thoroughly.

As in the case of the goat, introduction of the domesticated sheep into many countries was prehistoric. But in the case of sheep it was often followed by domestication, and then cross-breeding, of a local wild sheep. Sheep which were probably domestic (called turbary sheep) and descended from *Ovis aries palustris* were kept by the Neolithic Swiss Lake-Dweller people; such sheep may still be found in the canton of Grisons, and related races and types elsewhere in Switzerland and southern Germany.[1] One of these, the so-called Zachelschaf sheep, is common in Hungary; but, what is more interesting, it was common in Mesopotamia in the fourth millennium B.C. It had at that time already acquired

[1] Zeuner (1963).

its woolly coat, a character for which selection and breeding are necessary, since the wild sheep is not woolly but hairy, like a goat (*see* below). Another wild sheep which has contributed something to the making of domesticated sheep is the moufflon which is still found wild in Sardinia. Sheep first appeared in England with the Windmill Hill people.

But before going on to try to trace the line of domestication, something should be said about the most important of the changes which it brings about, the change from hairiness to woolliness.

Sheep were not domesticated for their wool. For, as I have said above, in the wild sheep that wool is not apparent, the animal is hairy; it has a short woolly undercoat obscured by a much longer hairy top coat.[1] The sheep, like the goat, was therefore domesticated first for meat only; later it became a milch animal. I suggest that, possibly as a result of the changes which occurred in domestication, the chance appearance of lambs in which the woolly undercoat outgrew the hairy top coat, led to the idea of trying to select and breed for wool to be used to make yarn, with the ultimate result of turning the sheep from a primarily meat-and-milk animal into a primarily fibre-bearing animal.

These changes which occur in the sheep's coat from the wild to the fully domesticated state—after a long passage of time, no doubt—are very remarkable. In the first place, all wild sheep have coloured coats, the wool is brown or reddish, never white; but in domestic breeds it is white, the pigment has been lost. In the second place, all wild sheep moult, shedding their coat every spring. It could, indeed, have been this fact which led to the breeding of sheep for wool; perhaps the first wool ever to be spun into yarn was not taken from the sheep's back, but collected from bramble and thorn bushes where wild sheep had rubbed off the moult-

[1] *See*, on this subject and conclusions to be drawn from the facts, Flannery (1969).

ing hair and wool. At all events, unlike wild sheep, domestic sheep do not moult, doubtless because, once shears had been invented, or some means of shearing, selection was against fleece-shedding which entailed loss.[1]

In his discussion of the history of sheep, Zeuner has light to throw on the ancient priority of goats over sheep in Neolithic times. It is not, he says, simply a question of goats being domesticated first and sheep later—although I must say that that seems to have been the order in south-west Asia. The primitive farmer arriving in new country, which was probably covered with bush and very often forested, preferred goats to sheep; for whereas the latter must have grass, goats are leaf-eaters and, in the long run, tree-killers. Their worst fault in our conservationist eyes was their greatest merit in the eyes of the primitive farmer, for they helped him to clear land of bush and trees. But in so doing the goat was eating itself out of a job, for once the trees were gone the grass could grow, and at that point sheep became preferable to goats. 'The order of the goat followed by the sheep is thus ecological and need not indicate that the goat was domesticated before the sheep. In steppe country the sheep is the obvious animal to be domesticated.'[2]

Now to consider the sheep in one of the two most ancient centres of agricultural and pastoral technology: Egypt.

There is very good pictorial evidence for the presence of the domesticated sheep in Egypt not later than the early fourth millennium B.C. Sheep depicted on the Nagada Palette are clearly recognizable as the screw-horned hair-sheep, a descendant of the Iranian Urial sheep and familiar during the fourth millennium in Mesopotamia. So domesticated sheep reached Egypt in Gerzean times.[3] There is also evidence of a kind—bones—that Asiatic sheep had reached Egypt even earlier, perhaps in the fifth, possibly even the

[1] Ryder (1969).
[2] Zeuner (1963).
[3] Zeuner (1963).

sixth, millennium. But here the trouble is of the kind indi-
cated at the beginning of this chapter: a case of sorting the
sheep from the goats, which Reed (1960) rightly insists must,
but often cannot, be done. However, there are bones for
periods earlier than Gerzean in an Amratian cultural context
at Abydos in Upper Egypt, in a Badourian context elsewhere, and on several Halafian sites. In no case can these
bones be certainly distinguished as those of sheep and not of
goats.

But there is a form of evidence for sheep-herding which
is unmistakable: wool. Where an archaeologist finds wool he
knows that sheep were not only present but must have been
quite old in domestication, since the coat change from hairy
to woolly has taken place. This kind of evidence has apparently been found in a Gerzean context at al-Omari.[1] But it
is at this same Gerzean period that evidence from bones does
clearly sort the sheep from the goats. The conclusion seems
to be that while Egypt may have had hair sheep *before* 4000
B.C., she certainly did have woolly sheep of Asian origin not
later than 3500 B.C.

So we are back to western Asia in the fourth millennium,
say about 6,000 years ago or rather less. And the pointer from
Egypt is fully confirmed: domesticated sheep were present in
Neolithic pottery communities [2] in Iran, Iraq and Turkestan.
The Hassuna period, named after a key archaeological site
and divided into phases, is clearly one of sheep fully domesticated: and this is so for the early Hassuna phases—5000–
4500 B.C.—as well as for the later. At the famous Tepe Sialk
in Iran, which is dated between 5000 and 4500 B.C., Vaufrey
found sheep bones in the lowest levels.[3]

The range of the Urial sheep in the wild was doubtless

[1] Reed (1960).
[2] Archaeologists divide Neolithic into pre- and post-pottery,
i.e. periods before and after a pottery industry in that context and
place.
[3] Vaufrey (1939).

wide, but it seems to have centred on Turkestan. However, it certainly extended to Iran and Iraq. I have already mentioned the site at Jarmo in Iraq, where bones of domesticated goats were found. Some specialists are satisfied that bones belonging to domesticated sheep can also be distinguished there. The Belt Cave in Iran, near the south coast of the Caspian, has also been referred to in connection with goats, and there also the excavator believed that he had found bones of domesticated sheep.

This Belt Cave had been occupied during three periods: in the most recent of these, later than 5000 B.C., by a Neolithic people who had pottery. Over 21 per cent of the bones found in the appropriate layers were those of sheep. In the middle layers, corresponding to a time between 6000 and 5000 B.C. and a pre-pottery Neolithic culture, 36 per cent of all the bones found were sheep bones. In the earlier layers below these, corresponding to a people practising a Mesolithic industry, only 2.7 per cent of the bones found were sheep bones, and below this, in an early Mesolithic phase, there were no sheep bones at all.[1] Coon argued from this that the domestication of sheep began in early pre-pottery Neolithic in Iran; that is, a changeover from merely hunting the wild Urial to penning or herding it. The date would be about 6000 B.C. To this Zeuner added that the 'control' of sheep may well have been even earlier in the lowlands of Turkestan where Urials were abundant in the plains.[2]

[1] Coon (1951).
[2] It should be said that Reed (1960) does not believe the Belt Cave finds to be satisfactory evidence for domestication of sheep.

6

Pigs

As anyone knows who has ever had anything to do with pig-keeping, pigs are clever animals. They are not only intelligent, they are extremely adaptable and highly emotional, capable of affection for human beings and even of devotion. George Orwell was quite right to make the pigs the leaders of his farmyard revolution against man,[1] for of all the denizens of the farm they are second in brain-power only to man; even the dogs are not as capable of original thought. Gilbert White (1789) tells of a sow which, when she had occasion for the services of a boar, made her own way to a neighbouring farm where a stud-boar was kept, opening a number of gates on her way by lifting the wooden latches with her snout. In more recent times there is the case, described in the magazine *Life* some years ago, of a French innkeeper who, having caught a wild piglet in the neighbourhood, reared it at home, until it became so attached to him that even as a fully grown boar it followed him everywhere, in the town as well as in the country. An acquaintance of my own, a successful pig-keeper, had a sow so in love with him that while she would tolerate male visitors accompanying her master to the sties, she would at once attack any woman whom he was careless enough to bring with him to see his pigs. When Dr Charles Reed was doing palaeozoological work at the archaeological site of Jarmo in Iraq he and his colleagues adopted some baby wild pigs—the pig is still wild in the neighbourhood—and kept them in camp. Reed (1960) says that they are appealing little animals which tame easily.

[1] *Animal Farm.*

In character and power of personality the pig is nearer to the dog than any other domestic animal. It has been of service to man in many dog-like ways, as well as in ways which are beyond the dog, though it is true that in some American cultures and in China dogs were, like pigs, bred for the table. In Périgord, the French centre for truffles, pigs were long kept as finders and diggers of truffles. The boars, perhaps, were not sufficiently tractable for this purpose, for sows were used for the work; and the truffling-sow was often accompanied by her piglets who thus learnt their trade. The sows were muzzled so that they could not eat the truffles which they found; apparently the Biblical prohibition of muzzling 'the ox when he treadeth out the corn' did not apply to pigs.

Zeuner (1963) has a fascinating account of the use of pigs as retrievers. When William the Conqueror made the New Forest a royal game reserve in the eleventh century, he prohibited the keeping of dogs above a certain small size; this was meant to keep down hunting, which he had made illegal for the wretched Saxon peasantry. The following account is from Kenchington (1949).

If small enough to creep through a large antique iron stirrup, now kept at the Verderer's hall in Lyndhurst as an interesting relic, dogs were considered too small to hunt. The dimensions in the opening of this stirrup are ten and a half inches by seven and a half. If you wished, for personal protection or other reason, to keep a larger dog, you could do so only if he were expedited. This meant that its forefeet were maimed by cutting off certain toes so that the deer stood in no danger from its sorties.

So, being unable to keep sufficiently large hunting dogs, the commoners trained pigs to help them in their hunting; and it seems that hunting pigs were in use well into and even later than the fifteenth century. In fact there is a nineteenth-century case of a retrieving 'Slut', the property of Sir Henry Mildmay, whose hunting skill is described as follows:

Of this most extraordinary Animal will be here stated a
short History, to the Veracity of which there are hundreds
of living Witnesses. Slut was bred in the New Forest and
trained by Mr Richard Toomer and Mr Edward Toomer,
to find, point, and retrieve Game as well as the best Pointer;
her nose was superior to any Pointer they ever possessed,
and no two men in England had better. Slut has stood Part-
ridges, Black-game, Pheasants, Snipe and Rabbits in the
same day, but was never known to point a Hare. When called
to go out Shooting, she would come home off the forest at
full stretch, and be as elevated as a Dog upon being shown
the Gun.[1]

In short, the pig is one of those animals which are psycho-
logically preadapted to domestication. This has a bearing on
our subject, for in the case of such an animal, and one with
such an enormous range in the wild, there is not much point
in looking for a single, primal domestication. The pig was
everywhere so readily available, and so predisposed to co-
operate with man, that it was domesticated over and over
again and in a great many different places.

There are only a very few species of wild pig with which
we need to concern ourselves. *Sus scrofa*, the most important,
has an enormous range; it is, or was, native to the whole
Mediterranean basin, all Europe, all south-western Asia and
parts of east Asia—in short, to the entire region of the most
ancient civilizations west of China. There is a north Indian
pig, native to Nepal, Sikkim and Bhutan, which may have
contributed genes to some races of domesticated pig. The
Chinese pig is called *Sus cristatus* and was domesticated separately; it is important to us because genes of the domesticated
Chinese pig now occur in our own domesticated breeds.
Many other names will be met with in the authorities, but
most of them have now lost their standing, and some refer
solely to domestic races.

From almost the beginning of its career as one of our close
associates in civilization, the pig has been divided into two

[1] From *Rural Sports*, vol. 13, 1807, quoted by Zeuner (1963).

quite distinct groups of domestic pigs. One is a large and rather long-legged animal, developed for herding; the other is a smaller, shorter-legged and more compact beast, intended for keeping in a sty. The extreme case of the small pig is the Chinese house-pig.

A pattern of domestication which can be traced archaeologically in the Neolithic Swiss Lake Dwelling sites is repeated in many other parts of the world. The local wild pig, hunted since Palaeolithic times, was still hunted; then a small, foreign, domestic pig was introduced from some culture farther east, but not necessarily from farther Asia as used to be claimed; then, as if this introduction set an obvious example, the local wild pig was domesticated. Late in the Neolithic, in fact, the Lake Dwellers had no fewer than three kinds of pig: a domestic local pig; a small pig which was introduced from the East; and a very small animal which may have been a house-pig, living with the family in their huts. Although crosses between these three breeds must have occurred from time to time, the three distinct breeds can still be distinguished by their remains.

We can come to a conclusion about at least one kind of prehistoric pig: some time between the early and late New Stone Age, European man domesticated the native wild pig and kept it in herds. So the honourable profession of swineherd is a very ancient one, only less ancient than that of shepherd and goatherd. For whereas the goat, and even the sheep, may be pre-agricultural in a few places, the pig never is, because his association with man could begin only after man had the means to settle down in one place and stay there. Consequently he is associated with farming and not with pastoral cultures; pigs are extremely difficult to drive and cannot, like sheep and cattle, be moved over great distances in herds.

For a long time it was believed, owing to certain details of the structure of the skull, that the small (turbary) pig was domesticated only in south-east Asia and was introduced elsewhere throughout its range in domestication, including south-west Asia and Europe. Domestic pigs with what were thought to be purely east Asian attributes have turned up in excavated sites which were the homes of the Körös people in Hungary—2500–2400 B.C.; in the Troy I level at Kissarlik; in the Neolithic B level at Sesklo in Greece, and on sites of even earlier cultures in Greece and Crete. Pig bones of this kind, survivals of still earlier settlements, were found at Ljubljana in Yugoslavia. Now there is a south-east European pig very like the south-east Asian animal; it is far more likely that the local pig was domesticated in south-east Europe, and distributed west into the rest of Europe, than that the small sty-pig and the house-pig came all the way from China.

Before we say anything about the prehistoric pig in the lands of ancient Semitic civilization, something had better be said about the curious and irrational tabus attached to this splendid animal. As everyone knows, pig-meat is tabu to Jews and Muslims; but it is almost certainly a mistake to think that this has anything to do with 'race'. Jews and Muslims bar pork, not because they are Semites but because they are peoples who take their laws from the Pentateuch. It is even surprising that strict Protestants, with their reverence for the Old Testament, consent to eat pork and bacon and ham. Formerly there were a number of non-Semitic people who held pork to be tabu; the tabu, however, does not seem to have applied to whole nations or sects, but rather to certain social classes.

Zeuner (1963) has shown very clearly that none of the usual reasons put forward in an effort to rationalize this tabu will stand examination. For example, it is not true that pork is unwholesome in hot climates: it is eaten with the normal measure of impunity in places such as New Guinea, the South Sea Islands, the Sunda Islands and, of course, southern China where it is the staple meat food. And as for the diseases incident to eating pig meat in certain circumstances, *trichinosis* being the worst, it is in the highest degree improbable

that a tribe of primitive nomads, or even the more sophisti-
cated ancients, would have connected such sickness with the
eating of this or that kind of food. To argue on those lines is
to credit the peoples of 1500–1000 B.C. with modern biological
and medical knowledge. Nor is there the least shadow of
evidence that there is any foundation for the tale, first advanced by Tacitus, that the Israelites barred pork because it
caused some kind of leprosy. As to that story, it has not, I
believe, been suggested by any authority (and I therefore
advance the suggestion with diffidence) that the threat of
leprosy may have been used by the Jewish religious establishment simply in order to discourage the eating of pork,
which was abhorred on grounds other than rational. That
would explain the passage in Tacitus; and the rabbinate
always had a regrettable tendency, of which they have not
yet cured themselves, to lay down the law as if they were
addressing children. They had, of course, to discourage pork-
eating not for any rational cause, but quite simply because
the Book forbade it. As for the Muslims, again there is no
problem: they too are people of the Book, revering the
Patriarchal laws. I know of no evidence that the pre-Islamic
Semites anywhere were forbidden pork.

   And that is why I find it difficult to accept a most interesting theory put forward by Antonius (1918) to explain this
tabu. He argued that because the pig was only useful to, or
even possible for, the sedentary farmer, therefore the nomads
—cattle people and horse people—who have always despised
the farmer, also despised his pig. Now it is certainly true that
for at least 7,000 years there has been an antithesis; between
on the one hand nomad, especially mounted nomad, male,
patriarchal, worshipping male deities; and on the other hand
farmer, dependent chiefly on cereals, female, matriarchal,
worshipping female deities. I tried in another book [1] to find
an origin for this antithesis, and there is a great deal to the
point about it in the writings of Robert Graves. But did the

[1] Hyams (1952).

nomads really carry this war to the point of not eating pork?
And did the non-Semitic Asian nomads have the same tabu?
I think not. There are, however, examples of food tabus
derived from hostility between ethnic or cultural groups. I
have read that the Chinese considered dairy produce unclean
because the Huns lived on it, and there was a time when the
upper-class English barred fish and chips, while to make
chips acceptable above a certain social level they have to be
called 'French Fried'.

It is quite likely that the Jews picked up their disgust of
pork in Egypt, during the Captivity. For the history of the
pig in ancient Egypt is very curious; although the picture
we can form is confused, it does seem that some ancient
Egyptians were barred by a tabu from eating pig meat; that
others could eat it only on certain days; that some of them
regarded mere contact with pig as a pollution requiring ritual
cleansing; and that in Egypt the pig was valued rather as a
working farm animal than for its meat. If you walk pigs, with
their dainty pointed hooves, over a field of mud left by an
ebbing flood—the Nile flood in this case—then the seed you
broadcast behind the pigs will settle into the earth at the right
depth. That is how the Egyptians used pigs—to help them
sow the corn.

Like all domestic animals in Egypt, the pig was sacred to
a god—in this case to an evil one, Seth. These things being
so, I believe that the Jews picked up their absurd tabu of
pork in Egypt during the Captivity, between c. 1600 and
c. 1300 B.C. And it is worth noting that the peoples who had
their pigs directly from south-west Asia or farther east, and
not by way of Egypt—the Greeks, for instance, and the Celts
—never had any objection to pork excepting when a religious
obligation *derived from an Egyptian source* was in question.

Pigs first appeared in Egypt in Neolithic times and had
become very common there by the third Dynasty—c. 2500
B.C. That dynasty seems, indeed, to have marked the zenith
of the pig's career in Egypt, for thereafter it became less

important.[1] It is from that period that we have the first written account of pigs. Earlier records are only archaeological, and Reed (1960) has rejected all the arguments for domestication based merely on bone-finds in Egypt, on the grounds that the pig was wild there and that the bones could just as well have been those of wild as of domestic pigs. An exception is made, with reservations, for a Gerzean site at Tukh. At all events, before 2500 B.C. the domestic pig was well established in Egypt.

Egypt was not the only country where the pig was domesticated or introduced no less for its value to the farmer as a working animal than as a source of meat. In forested Europe it was invaluable in the clearing and preparation of land for arable farming. I can best demonstrate how by telling a story from personal experience. In the 1950s an acquaintance of mine was able to buy a field very cheaply because, having been used as a silver-fox farm, it had stout wire netting buried quite deeply under the surface of the soil. This netting had, no doubt, originally been laid on top, but had been trodden and grown in. It was thought that the cost of removing it would be high, so its presence reduced the value of the land. My acquaintance put a small herd of pigs onto it, and they had that wire up and lying loose in a very short time. This rooting power, as it were, of herds of pigs was discovered a great while ago. Such a herd working through a forest will destroy the undergrowth, prevent tree regeneration by eating such tree seeds as beech nuts, acorns and fruits, destroy such pests of crops as snails and slugs and mice. So from Greece to Britain in the late Neolithic and Early Bronze Age the pig was as much a kind of living agricultural machine as a source of meat.

The pig was first domesticated in Mesopotamia, but the earliest remains, which even the most cautious and sceptical of experts admit as those of domestic and not wild pigs, are not as early as those of goats or sheep. Bones which are

[1] Zeuner (1963).

unmistakably those of domestic pigs have been found at Anau and Shah Tepe,[1] and the significance of those place-names is that by 3500 B.C. at latest the farmers of the Meso-potamian cultures were keeping and breeding pigs. Is there any evidence for a still earlier domestication? Pig-bones were found in the Belt Cave, but only at post-pottery levels. Pre-pottery levels at Jericho yielded only bones of wild pig.

A figurine carved in ivory of a clearly domestic pig—very fat indeed, naked, not bristly, and with drooping ears—was found at Tell Agra, in Sumeria, and dated between 2700 and 2500 B.C. Now, a pig so enormously different from the wild animal could hardly have been evolved in only a few generations; centuries of domestication are implied in its mere appearance. So this serves to reinforce the evidence of the bones—domestication in the fourth millennium B.C.

Until early in the nineteenth century the domestic pig of south-west Asia and Europe, and therefore of America, owed nothing to any species but *Sus scrofa* and its sub-species. But then the shape of the domestic pig began—with the English 'Berkshires', England being the leading country in the Agricultural Revolution which preceded the Industrial Revolution—to change with the introduction into breeding programmes of the Chinese pig, *Sus vittatus*. This Chinese pig, smaller, stockier and quicker to fatten than our own, had for thousands of years been by far the most important source of meat in China. But for exactly how long? It seems that we do not know. The southern Chinese domesticated a species called *Sus cristatus*, while the northern had a wild pig which seems to have been intermediate between *S. cristatus* and *S. scrofa*. Domestication no doubt goes back to the Neolithic in China, as elsewhere, but there is no evidence to support an old idea, now rejected by most authorities, that it was from southern China that the domestic pig spread all over Eurasia. It is, in fact, unlikely that the Chinese had domesti-cated a pig as early as the Mesopotamians.

[1] Reed (1960).

7

Cattle

Were it not for the importance of wool, one would say that cows are the most important of all domestic animals. Still widely used for farm traction, for many centuries cows and oxen alone drew the plough and the cart, being employed for that purpose long before the horse was thought of. There is no need to labour the point of the importance of milk, beef and hides in the world's economy. Even the hoof and horn are very valuable in commercial horticulture for fertilizers, just as in prehistoric times they were valuable as industrial raw materials. And cattle are so important to some primitive peoples that they can be said to live parasitically on their herds: the Masai, for example, who draw not only milk from their cattle but also blood to drink. Whole cultures have been shaped by the nomadic nature of a cattle economy: the male-dominated, patriarchal, sun-or-sky-worshipping cultures as opposed to the sedentary, female-orientated, matriarchal, earth-worshipping cereal cultures. In some low-rainfall parts of the world—large areas of Argentina, for example, and even larger areas of Australia—greed in overexploiting cattle as converters of grass into meat and hides has led to the ruin of the land; but we cannot blame the animal for that.

All the peoples of Europe and Asia had domestic cattle before the beginning of history. As the peoples of Europe begin to emerge out of prehistory, with specific names and attributes—Greeks, Celts, Latins, Teutons—they are already accompanied by their herds of cattle, and in some cases are living on those herds rather than on vegetable crops.

I have several times referred to the Swiss Lake Dweller people, the folk who brought with them Neolithic industrial and agricultural techniques when, having migrated from somewhere farther east, they settled in communities on some of the Swiss and Italian lakes, building houses on stilts over the water. Although relatively advanced they did not yet have the use of metals: but they already had herds of cattle.

In 1862 L. Rütimeyer made a study of the animal bones on these Swiss Lake Dweller sites.[1] At first he distinguished three breeds of domestic cattle, calling them *Bos primigenius*, *Bos brachyceros* and *Bos trochoceros*. But later, Rütimeyer came to the conclusion that the animals he had called *trochoceros* were only variants of *primigenius* cattle. So we are left with two prehistoric breeds of domestic cattle: the *primigenius* and *brachyceros*. The latter is now more commonly known as *Bos longifrons*, so that is the name I shall use.[2]

They were quite distinct: *primigenius* was a big, long-horned beast—one can still see his obvious descendants in the lovely lyre-horned, grey traction-cattle of Tuscany; *longifrons* was a smaller beast, with short, inward-curving horns and the high forehead to which it owes its specific name. On some sites only *longifrons* remains were found; for example, in the lowest, that is earliest, Neolithic level at

[1] Rütimeyer (1862).

[2] The subject is greatly complicated by a factor which I am bound to mention. It is called by zoologists 'sexual dimorphism', by which they refer to the morphological differences between male and female of the same species. In the genus *Bos* the male is larger than the female. It is therefore argued by some authorities that the small skulls and other bones of cattle found on Neolithic sites are those of cows, the big ones of bulls. The best exposition of this argument known to me is by Caroline Grigson (1969). The subject is also discussed by Zeuner (1963). It seems to me, however, that the evidence is weightier on the side of decrease in size of *B. primigenius* in domestication, and I have written this chapter in that sense.

Saint-Aubon on Lake Bienne. At other sites both kinds of remains were found. At Seematte-Gilfingen, on Lake Zürich, and elsewhere there were three kinds, readily distinguishable by size. The third was intermediate between *primigenius* and *longifrons*, and can best be explained as a hybrid between the two; indeed, such a hybrid would have been bound to occur where both kinds of cattle were herded in primitive stock-breeding conditions.[1]

The three different sizes of cattle, with one or two morphological differences as well, have been found on Neolithic, Bronze Age and early Iron Age sites elsewhere in Europe and over a very wide range of territory. But there is no real point in bothering with other sites; for our purpose the Swiss Lake sites will do very well.

Leaving the small *longifrons* cattle aside for a moment, there is no difficulty in deciding where the big, long-horned cattle came from: they were natives.

Since early Pleistocene times a species of wild cattle, *Bos primigenius*, had roamed and grazed all over Europe. The correct vernacular name is aurochs.[2] For many thousands of years this aurochs was an important game animal for Palaeolithic man, witness the magnificent paintings of cows and bulls made by the Cro-Magnon artists who inhabited the caves of Lascaux and Les Eyzies in the Dordogne valley between 30,000 and 15,000 B.C.

The *primigenius* breed of domestic cattle, whose remains have been found at Swiss Lake Dweller sites and elsewhere at an early Neolithic level, are domesticated aurochs; and there is no doubt that they were domesticated in Europe either in one centre or in several. The only question is, when? There seem to be two possibilities. One is that the arrival in Europe of a breed of domestic cattle—*Bos longifrons*—from the East suggested or stimulated the domestication of the

---

[1] Zeuner (1963).
[2] Sometimes, but incorrectly, used to mean bison, buffalo and other bovids.

native and local aurochs; the other is that the aurochs was already domesticated when migrant people bringing *longifrons* with them arrived in Europe.

I have elsewhere tried to 'reconstruct' the kind of events which might easily have led to the first domestication of ruminants and horses: the rounding up and corralling of small herds so as to keep meat 'on the hoof'; the discovery that the animals, especially the calves, foals, lambs and kids, were by no means intractable; and the realization that, as a result of the animals' crop-robbing propensities, it might be possible to maintain a more or less permanent herd (even though only a mere nucleus of it over-wintered in the colder places) by keeping it under control and so forcing the animals to breed in captivity. In the north temperate zone this last technique was not perfected for 5,000 years; not until the development, in the seventeenth century, of fodder crops, especially turnips, which could be kept as winter feed.

The wild aurochs has long been extinct; but a curious, interesting and successful attempt to 'reconstitute' it was made in Germany, and is worth a digression if only because it establishes beyond doubt the origin of our domestic cattle. Two German zoologists, Drs Lutz and Heinz Heck, were the men responsible. In 1921 they began two experiments in 'breeding back' to the wild aurochs, using chiefly breeds which showed clear aurochs characteristics. It is a known fact that all the genetical attributes of the ancestors of an animal, or of a man for that matter, remain everlastingly potential in the genes; characters which have, by selection, segregation, etc., become recessive can be brought out again if combined, on mating, with the same recessive character in the other parent. (It is because certain physiological flaws, even madness, are genetical attributes that the mating of close kin in man has been discouraged, and that the remarkable psychological barrier against incest has been built up.[1]) So what

[1] In this connection it is necessary to point out that in animal breeding, and in human social rules derived from the observed

the Hecks were trying to do was to 'recombine' the gene
which governed the typical aurochs characters.

Heinz Heck, who was Director of the Munich Zoo, worked
with Scottish Highland cattle, some Alpine breeds, Frie-
sians, and Corsican cattle. In due course he was rewarded
with one male and one female calf which had 'Lascaux'
characteristics, as it were. These were mated; their calves
'bred true', and by 1951 Heinz Heck had about forty recon-
stituted aurochs.[1] Lutz Heck, Director of the Berlin Zoo,
used Camargue and Spanish cattle bred for the bull-ring,
and primitive breeds of English park cattle. He too was
successful, and although most of his aurochs were lost in the
bombing of Berlin, some of his stock is now running wild
in the Białowieza forest in Poland, where also the last
European bison are preserved.[2]

Those experiments were not a mere scientific *tour de
force*: they proved the origin of the modern domestic breeds
of cattle in the aurochs of the Pleistocene. So there is no
doubt about the origin of the big long-horns. The case of the
smaller short-horns is not so clear, even though the Hecks
used some modern short-horn breeds.

The only plausible explanation of the presence of *longifrons* in prehistoric Europe is that they were brought from
the East: those immigrants who settled round the Swiss
and Italian lakes brought the breed with them. But where,
originally, did those cattle come from? It will be best to take
a look at two parts of the world where the Neolithic revolution was in advance of the rest of the world.

It is difficult to say just what happened in Egypt: the

---

facts of heredity, practice long preceded theory. Now that we have
a science of genetics we are able to plan breeding programmes with
more chances of reaching a predetermined result; but the want of
that science did not, demonstrably, prevent empiricists from getting
results.
[1] Heck, L. (1951).
[2] Zeuner (1963).

picture is at best confused. Cattle bones have been found in the Neolithic A levels at Kom, in the Fayyum, and radio-carbon dated to between 4500 and 4100 B.C. Zeuner says that these were the bones of wild cattle, while Reed appears uncertain of what they were. In Egyptian Neolithic most of the bones found seem to have been hard to identify, and at the early levels at which they can first be reliably identified as domestic the dating becomes uncertain. The earliest dates for bones of unmistakably domestic cattle are for the pre-dynastic Amratian period—3700 B.C. by radio-carbon.

There is evidence other than bones for a date only a little later: a charming little clay model of *Bos taurus* (i.e. domesti-cated *Bos primigenius*), which was found in a grave at El Amrah; and other models of cattle from graves of the pre-Dynastic Nagada period, say about 3500 B.C. It looks rather as if the pattern might have been like the European one but earlier: introduction of *longifrons* cattle, followed by domesti-cation of the wild *primigenius* of the Nile Valley. But that is a guess; the two distinguishable domestic breeds of cattle in Egypt may both have descended from the native wild aurochs.

Once we reach the Dynastic period we have both figura-tive art and dynastic texts upon which to draw, and by then four domestic breeds can be distinguished. Two are of the *Bos primigenius* stock, but with different horns—the true double lyre-shaped horns have emerged in one; the third is a *longifrons*-type short-horn; the fourth a breed of hornless cattle. Conceivably they could all be descendants of *Bos primigenius*, though it seems more likely that the short-horns and hornless kinds came from stock introduced from Mesopotamia by way of Palestine. There is, however, no unmistakably alien character in Egyptian cattle until the Eighteenth Dynasty, when humped cattle, which can only have come from India *via* Mesopotamia and Palestine, make their appearance.

It goes without saying that the cattle of Egypt became sacred. The Apis Bull and the Cow Hathor are among the most celebrated of man's millions of gods. But this apotheosis, however ancient its origin, is a later growth. The earliest known written account of the Apis Bull is from the reign of King Khasekemui of the Second Dynasty; it is engraved on what is known as the Palermo Stone, dated *c.* 2800 B.C. There are Hathor-like headings to the Narmer Palette, and some very early *bucrania* [1] shown decorating the mastheads of ships painted on Nagada II pottery. That takes us back to about 3200 B.C.; but the art style is not Egyptian, it is indeed very alien, probably Mesopotamian.

Not even in Egypt, however, was the Apis cult as important as the bull-cult in Minoan Crete. The bulls used for the lovely bull-leaping ballet must surely have been specially bred for the ring, just as fighting bulls are bred today for the Spanish ring. It is true that a number of respectable authorities have argued that the bulls of the Minoan arena were wild animals, but I do not believe it. It is not likely that there ever were truly wild bulls on an island as far from the mainland as Crete; nor is it likely that any wild bull would have had the piebald markings depicted in some of the pictures of bull-leaping; [2] and it is most unlikely that the feats performed by the bull-leaping team of dancers could have been performed with animals which had not been trained, and maybe even bred, for the work.

Zeuner suggests that in about 3500 B.C. cattle-breeding immigrants reached Crete with domesticated cattle, presumably from Asia Minor; that some of these cattle became feral; and that it was from the resultant population of naturalized cattle that the bulls for the arena sports and for sacrifice were obtained. But that was all later: the Minoan cups at Vapheio, near Sparta, showing the capture of bulls with nets and a decoy cow, and the hobbling of the captive bull, are dated 1500 B.C.

Professor Max Mallowan (1956) believes that this Minoan

[1] A *bucranium* is a bull's head used as a symbol.
[2] Zeuner (1963).

bull-cult was of west Asian origin. The *bucranium* symbol
was being used in northern Iraq as early as 4500 B.C., and
among his finds were double-axes—a symbol of divinity in
Crete—associated with bulls' heads:

In the Dictean cave on Crete, where Zeus was supposed to
have been born, masses of double axes were found which
had evidently been used as sacrifices. The connection
between the double axe and the bull survived into much
later times. From the Hittite period, about 900 B.C., a god
holding a double-axe is known standing on a humped bull. The
figure of Jupiter Dolichenos standing on a bull and holding
the double-axe in one hand and a bundle of flashes of lightning in the other is evidently descended from it. This Roman
deity was venerated from Mesopotamia across Anatolia into
the Balkan peninsula, up the Danube and down the Rhine
into Britain. Professor Mallowan attached particular importance to the architectural parallels found in the circular
buildings at Arpachiah in Iraq and at Khirokitia in Cyprus.
He thus finds it hard to avoid the conclusion that none of
these things could have appeared in Crete in particular and
elsewhere in the Mediterranean 2,000 years later had they
not ultimately descended from the energetic villagers of
prehistoric Assyria.[1]

To which Zeuner adds that beings like the Minotaur, with
a bull's head on a man's body, are familiar in Babylonian
mythology.

Bull-leaping must have been extremely dangerous; it may
well have been a method of making human sacrifices to the
Minotaur, as well as a public spectacle. A dancer killed by the
bull might be thought of as a victim chosen by the bull. But
Zeuner suggests that it may have arisen from the rural art of
bull-catching, itself a merely necessary part of the business
of stock-breeding. In that connection, here is a short passage
from Mr John Yeomans's *The Scarce Australians* (1967),
about the handling of semi-feral and even feral cattle in the
outback of Australia's Northern Territory in our own time:

[1] Zeuner (1963) quoting Mallowan (1956).

As soon as the beast is sighted on such a hunt, a mad pursuit begins. The cow or bull may lumber away over boulders, around trees, down gullies and up hill-sides covered in loose pebbles, for eight or ten miles. Sometimes the beast will escape. Sometimes the horseman, near exhaustion himself, runs the beast to exhaustion. Now, as in another sort of bull fight, comes the time when the horses have no further place in the struggle and the man must face the beast alone. The beast stops and turns on his pursuer, the rider hurls himself off his horse, runs towards the panting beast and grabs it by its tufted tail. The man pulls and the beast swings its head round, hoping to hook the man with its horns. The man, pulling still harder on the tail, steps close to the animal's side and pushes with his shoulder. The cow or bull loses its balance and falls to the ground. The man instantly seizes one of its hind legs and pulls it straight up in the air. This stops the beast from scrambling to its feet. The man then grasps the rope or leather strap he had been carrying over his shoulder and round his waist and lashes the beast's hind legs together. The man then sits down on the ground at a safe distance and wonders if there is any easier way of earning a living. After that he climbs back on his horse and rides to the stock camp, where he collects four or five of the docile cattle and drives them back to the leg-tied beast.

The docile cattle are driven to form a loose mob surrounding the captured animal, from which the leg ties are soon removed. Finding itself with its own species, the wild cow or bull almost always allows itself to be driven with the others back to the stock camp where the captive is either put in a holding corral or guarded by a stockman on horseback.

In some such exercise as this did the art of bull-leaping have its origin: it was a stylization of an economic act, a ballet based on the cowboy's daily work; it was doubtless also a sacrificial rite.

So the earliest dates yet found for domestic cattle are west Asian, Mesopotamian. And since we are still seeking a source for the *longifrons* breed, which appears before or with the *primigenius* cattle in Neolithic contexts everywhere, surely this must be it. Two theories have been advanced to explain the origin of this breed: that there was another wild species,

12

12. Indian Jungle Fowl. The physical differences between the wild bird and the domesticated farmyard and battery fowl are not so great as to make recognition difficult. Selection and controlled breeding over thousands of years produced the 'unnatural' faculty of being able to lay up to three hundred eggs a year.

13

13. Wild Cat (*Felis sylvestris*) on the increase in the Scottish Highlands and moving south. Also native to the Balkans and West Asia.
Ancestral to the domestic cat, which was introduced to Britain as
such, but readily mates with its wild cousins.

14. Cheetah: a case of domestication never completed, it was used
by man as a hunting partner but was never transformed for man's
service.

15. Mongoose (*Herpestes nyula*), another case of incomplete
domestication: used before the cat to keep houses free of vermin in
India, it was never transformed into an animal permanently associated with man.

14

15

16

16. The Moufflon (*Ovis musimon*), now confined in the wild to the mountains of Corsica and Sardinia, is one of the ancestors of our domestic sheep.

17

18

17. The Soay sheep of the isolated St Kilda Islands still show recognizable Moufflon characteristics. The transformation of these goat-like animals into domestic sheep of the kind represented by the ewe and her lamb (18) above is even more striking than appears, for thousands of years of selection have eliminated the coarse hairs from the fleece to leave only the soft, fine wool.

19

19. Ibex is one of the wild goats still to be found in parts of Europe. Goats were probably domesticated before sheep for they not only provide meat and milk but, as browsers rather than grazers, are of great help in clearing woodland and scrub for agriculture, a virtue which becomes a vice where reafforestation is attempted.

20

20. Another ancestor of the domestic goat is this Wild Goat (*Capra hircus*) which still ranges through south-east Europe, Asia Minor and Persia, into Pakistan.

21

22

21. Feral Goats, not easily distinguished from truly wild goats, are thought to have returned to a wild way of life from domestication, which both goats and sheep do very readily in suitable country. Compare these goats with . . .

22. . . . this British Saanen, the fine flower of milch-goat domestication in which not only is the milk-yield much increased but the very bone structure is modified.

now extinct, smaller than *primigenius* and with shorter horns; or that *primigenius* is also the ancestor of *longifrons*.

There is no great difficulty about accepting the second theory. I have referred in other chapters to the morphological changes which follow domestication; and if, as seems to be the case, *primigenius* had been domesticated in western Asia many centuries before *longifrons* appears outside that very advanced region, in Neolithic contexts, there would have been time for some very considerable changes to occur. It seems that even in the wild state some *primigenius* populations may well have been smaller than others. Apparently if a population of such animals becomes isolated, say in some remote valley, or when an erstwhile peninsula becomes an island by reason of some change in sea level, or a land subsidence, one of the consequences of the resultant in-breeding can be decrease in size. Then there is what zoologists call Bergman's rule, that mammals decrease in size with increasing southerly latitude. Besides, when it comes to the initial acts of domestication, it would be natural for men to choose the smaller animals of a wild herd and those with shorter horns, in other words the less dangerous individuals, when cutting out beasts for coralling and perhaps taming.[1]

On the other hand, Reed (1960) has this to say:

In addition to the large wild cattle that are known from Farmo and shown in Assyrian hunting reliefs, there was also, at least in the northern mountains during the late Pleistocene, a population of small wild cattle hunted by the people who lived in the Shanidan cave . . . The cattle were surprisingly small and no remains of large cattle were found. I mention the Shanidan cattle only to indicate how little we really know of the details of the distribution of the wild populations to which we must look for the ancestry of our domestic cattle.

There is no evidence that the Chinese domesticated cattle in their own country; they had their cattle from the West,

[1] *See*, in this connection, Grigson (1969).

both the *primigenius* and the *longifrons* breeds. It is not clear when this introduction occurred, or exactly where from; but as, until comparatively recently, the Chinese used cattle only as working animals, and neither for meat nor for milk, it is perhaps unlikely that they had them from a northern source, for all the north-central Asian people make great use of dairy produce and have done so from a very remote epoch. Zeuner thinks that the Chinese must have had their cattle from a south-westerly, in fact from an Indo-Chinese or Indian, source, but before the invading Aryans introduced the use of milk into India, that is before 1500 B.C. This hypothesis is reinforced by the fact that very early Chinese cattle in the south of China show signs of the zebu hump.

Zebu are humped cattle with long, narrow faces, very erect horns, dewlaps, and drooping ears. They are as a rule confined to hot countries, notably India and tropical Africa. Their origin is uncertain, because although most of the archaeological evidence points to India, a good deal of it comes from Mesopotamia: a rough figurine found at Arpachiya in a Halafian context is dated *c.* 4500 B.C., but one wonders if the resemblance to zebu is accidental. Representations of humped cattle have been found at Nineveh and on New Kingdom sites in Egypt, but they do not become common in the eastern Mediterranean until very late, *c.* 600 B.C. The link between the Indus Valley civilization and Mesopotamia, and therefore Palestine, Syria and Egypt, is probably provided by the pictures of humped cattle on pottery found at Rama Ghindai in northern Baluchistan, in a Chalcolithic (fourth-millennium) context. But the most abundant evidence for prehistoric humped cattle is from Mohenjo-daro.

Those Indus Valley people, the farmers of the Mohenjo-daro and Harappa countryside, had both zebu cattle and the familiar *primigenius* (*Bos taurus*) breeds. The latter they must have had from West Africa by trade; the zebu they may have domesticated for themselves. Throughout the Pleistocene there was a species of wild cattle, *Bos namadicus*, in India;

and it seems to be the only possible source of an Indian domestic breed. Domestication must have been accomplished in the fourth millennium B.C.—I find that Halafian zebu rather difficult to believe in—and zebu cattle exported to south-west Asia, just as *primigenius* and *longifrons* cattle were imported from there, by way of Baluchistan.

Western Asia is, then, the primary source of our domesticated cattle, and above all of the *idea* of domesticating cattle, as of so many good things. No other part of the world has contributed so many of the living artefacts which are the very basis of all civilization.

But the kind of cattle familiar to us in the West is by no means the only kind which has been fully domesticated. A number of other species of the same or related genera have been brought under control in several parts of the world, and the resultant domesticated breeds are, for the most part, interfertile with our own.

There is a group of animals distributed all over the world belonging to the genera *Bos*, *Bubalus*, *Syncerus* and the subgenus *Bibos*—wild cattle, buffaloes, bison and the like. Some of these have been domesticated, the others never have; and although there must be reasons why some species have been neglected, they are not apparent, so that the choice of species to be domesticated looks like chance. In a sense, it was chance: for example, the buffalo put itself in the way of being brought under man's control by persistently raiding his plantations, whereas the bison is never a crop-robber.

But since there are other ways besides crop-robbing in which animals initiate some kind of social relationship, even though it be one of mutual hostility, with man, the mere fact that the bison was not a crop-robber would not explain why its domestication was never undertaken, were it not that the explanation is self-evident. Although by the time of the Discovery most, but not all, of the North Americans had reached the stage of primitive farming, and some were much more advanced along that line of development, probably none had reached the point at which the domestication of large and not very tractable animals, unsuitable for nomadism, becomes possible. And if that reason alone be thought inadequate, there is the fact that, with their maize patches and their rich hunting, they were not sufficiently 'challenged' by their environment to bestir themselves; culturally they were comfortably fossilized in some stage of Neolithic.

It is, on the whole, more curious that the domestication of the European bison was never, as far as we know, undertaken; of course, it may have been, and all trace of the effort lost. The animal would have been useful for a number of purposes, and it can hardly have been more intractable than the gaur of India and Assam (*see* page 92) which was long ago successfully domesticated, though it is one of the most ferocious animals in the world.

Again, whereas the Indian water-buffalo was domesticated thousands of years ago, the African buffalo never was. This is the more remarkable in that some north-west African peoples were able to capture and tame elephants—the elephants which, for example, the Carthaginians used in warfare.

THE YAK

A number of animals dealt with in this book belong to a particular category in that they enabled peoples living under special and usually harsh conditions, in the territories marginal to the high civilizations, to have the services of domestic animals although the superior ones already long since domesticated could not thrive in the places where those hardy peoples lived. Thus the reindeer was a horse-and-cow substitute for the peoples of the Arctic North, and the camel for those of the arid steppes. The yak, with its ability to thrive and work at enormous altitudes and in conditions of intense cold, is another case in point, exploited by the Mongolians and Tibetans. The yak is the Tibetan's ox, the

Mongolian highlander's cow, doing duty for superior kinds of bovine cattle which could not thrive in its habitat.

Although fly-whisks or swatters made of the long and silky-haired yak-tail were reaching European markets from the East or from Central Asia in the first century of our era (they are mentioned by the poet Martial, XIII, 71),[1] it is unlikely that any European had any idea what this animal was like until more than a thousand years later. In the course of his travels, during the second half of the thirteenth century, Marco Polo first came across yaks at a city which he calls Singui in Tangut, and which one of his editors [2] identifies as Si-ning in the western part of Shen-si, a great entrepôt town on the road from Lhasa to Peking and Shanghai. This city, north-west of Langchow and east of Lake Koko Nor and of Tsing Hai, is on the 3,000-metre contour (9,840 feet), just the sort of altitude to suit yaks which are, indeed, at home at much greater altitudes. Marco describes the yaks as follows:

Here are found many wild cattle that in point of size may be compared to elephants. Their colour is a mixture of white and black and they are very beautiful to the sight. The hair upon every part of their bodies lies down smooth, excepting upon the shoulder where it stands up the height of about three palms. This hair, or rather wool, is white, and more soft and delicate than silk . . . Many of these cattle have become domesticated and the breed produced between them and the common cow are noble animals and better qualified to undergo fatigue than any other kind. They are accustomed to perform twice the labour in husbandry that could be derived from the ordinary sort, being both active and powerful.

Writing, or rather dictating, many years after the event, Marco exaggerates the size of the yak, but otherwise his description is not bad. And perhaps even in this matter of size his exaggeration is not as great as one might judge when looking at a yak in a zoo, or at work on the borders of Tibet.

[1] Zeuner (1963).
[2] Everyman's Library, 1918.

It is not impossible that the yaks of Singui were much bigger than those with which we are familiar; after all, the Indian water-buffalo, bred for size in India, is gigantic, whereas the same animal, bred for tractability in south China, is very small. As to its strength, a modern yak can easily carry 300 lb. over rough mountain tracks at altitudes of 15,000 feet and more; and at such altitudes it is driven, ridden, milked and used as a pack animal. A point of importance in our context: Marco describes the yaks as piebald, a sure sign of ancient domestication.

The centre of the wild yak's range in nature is Tibet, extending far into Mongolia. It is therefore certain that either the Mongolians or the Tibetans accomplished its domestication. And since the silky yak-tail fly-swatters were on sale in Italian markets in the reign of Domitian, we can at least be sure that this domestication was accomplished before the birth of Christ. There is no date enabling us to decide how long before, but here are some speculations which may not be without interest.

Let us take Marco Polo's Singui as typical Yak country. As I have said, it was not very far from modern Tsing Hai, and Tsing Hai is one of the sites of a Palaeolithic microlith industry belonging to a culture whose type-site is Chow-ku-tien, excavated by Professor Pei Wenchong in 1954. So there were hunting peoples in this very high country long before any kind of husbandry, and it is obvious that, since the territory is well within the natural habitat of the wild yak, the yak must have been one of their quarry animals. And where there is a more or less continuous development of one culture into another, as Palaeolithic gives way to Neolithic, hunting to stock-raising, yesterday's game animal is to-morrow's domesticated animal. But although the challenge of very harsh conditions—extremes of heat and cold, aridity, extreme altitudes—does produce in human beings responses of extraordinary ingenuity, it is often excessive in the sense that the effort of response is exhausting, and consequently the culture in question is apt to be retarded. In short, people living at altitudes near 2,500 metres and more were unlikely, until late by comparison with lowland cultures, to attain the level at which it becomes possible to domesticate animals other than those which, like the dog, 'ask for it'.

The yak can, perhaps, hardly have been in domestication for fewer than 3,000 or more than 4,000 years.

THE BUFFALO

It was the opinion of Alexander von Humboldt that the Indian water-buffalo, still a useful farm animal in parts of Italy, was not introduced into Europe from Asia until the time of the Crusades. But Humboldt may have been mistaken. The first European account of the animal is Aristotle's, but he probably knew it only by hearsay; it is unlikely that he had actually seen a buffalo, for although it appears that Europe once had the same or a very similar animal as a native, it had been extinct in Europe for thousands of years before Aristotle's time.

There may, however, have been an introduction of the domesticated buffalo into Italy some centuries before the first Crusade. When, in A.D. 723, Saint Willibald saw Indian buffaloes in the ghor of the Jordan Valley, he was much surprised at their strange appearance; the implication is that he had never seen them before. Yet, 'as Keller [1] points out, St Willibald had travelled through Italy and Sicily where buffaloes are to be seen today, so they cannot have been there then.' [2] I do not think this follows. I am familiar with buffaloes in India and Indonesia, but having travelled extensively in all the provinces of Italy, I have yet to see a buffalo in that country, although I know that there are many. That Willibald did not see any does not mean that there were none.

[1] Keller (1913).
[2] Zeuner (1963).

The Lombard historian Paul the Deacon says in his *Historia Langobardorum* (*c.* 790) that buffaloes first arrived in Italy during the reign of the Lombard King Aigulf, and that they were regarded as wonders by the Italian people. Aigulf occupied the Lombard throne about A.D. 600. Victor Heyn (1888) suggested that the buffaloes might have been a gift to the Lombard king from the Khan of the Avars. The latter were then on the Danube, and the Lombards had sent them skilled shipbuilders, so the buffaloes might have been a return present. Keller dismisses this, but I do not know what good reasons he had for doing so. One thing remains certain: the water-buffalo reached Europe from India, whether it came in the seventh or the twelfth century, whether by way of Hungary or of Palestine.

The animal here in question is the *arnee*, the domesticated version of *Bubalus bubalis* which in the wild state is confined to India and Ceylon.[1] It lives always near water, for it must be able to wallow; not even the domestic arnee can dispense with this indulgence. In other respects buffaloes have the same wants and serve the same purposes as the ordinary kinds of bovine cattle—draught, burden and milking—and for those purposes were domesticated in India at a very remote epoch.

Some animals owe their servile state to their temerity in stealing man's crops, or perhaps even to competing with him for edible plants in the wild. It is probable that buffaloes were first captured by man, perhaps by trapping, when they took to raiding the gardens or plantations of Neolithic farmers; and the question remaining therefore is how early in our history did this kind of accident result in domestication.

The late Henri Frankfort (1939) first illustrated a cylinder seal, 'the seal of the servant of Sargon' of the Akkadian dynasty; the design shows two buffaloes as what heralds call 'supporters'. There are other examples of such seals, all of

[1] Zeuner (1963).

them dated about 2000 B.C. or a shade earlier. Some authors account for this by claiming that the Indian buffalo was formerly found wild in Mesopotamia; but there is no palae-ontological evidence for this, and it is much more likely that the animal was introduced as a domesticated species, from India. The earliest evidence we have for its presence between Tigris and Euphrates is a seal impression found below the level of the royal cemetery at Ur, and which was therefore made prior to 2500 B.C. The obvious conclusion is that the domesticated buffalo was introduced into Mesopotamia in the first half of the third millennium B.C.

It might seem odd, if that be so, that it took at least 3,000 years for the beast to reach Europe; but it is not so odd as it seems. In the first place, the Near and Middle East cultures were for thousands of years far in advance of the European; in the second place, the climatic conditions and the dry agri-culture of most of Europe made the buffalo less useful there than in the East. But it really is remarkable that the animal did not reach the Nile Valley until well into the Christian era, for as Zeuner points out, it would have been most valu-able there. The dissemination eastward was much faster, and again for obvious reasons: the eastern farming cultures were more advanced, and the chief use of the water-buffalo is in the cultivation of 'wet' rice. This, until recent times, was confined to south-east Asia. The buffalo was well established in Indonesia and south China long before the beginning of the Christian era. But for really early dates in countries to which buffaloes were introduced, we cannot do better than refer to Mesopotamia.

If the introduction was accomplished there, as we have seen that it was, before 2500 B.C., then obviously the buffalo had been brought under man's control long before that date in its native land. But as far as I know the earliest actual evidence we have is again a seal impression (date about 2500 B.C.), from Mohenjo-daro in the Indus Valley (Harappan) Culture. Zeuner (1963) illustrates this as showing that domestication had been accomplished by that date, although in the accompanying text he says that this is not certain, since the animal depicted could have been a wild buffalo. The weight of evidence, however, seems to me to be that domestication had, in fact, been accomplished in India early in the third millennium B.C. And since the Indus Valley civilization was the most advanced in India at that time and had, moreover, trading relations with the Meso-potamian city-states, then it was surely the farmers of that culture who domesticated *Bubalus bubalis*.

GAUR AND BANTENG

Bos (*Bibos*) *banteng* is a kind of wild cattle native to Java, Borneo and Burma. The banteng bull has a reputation for extraordinary savagery, and travellers in the Indonesian bush have alarming stories of its instant, lightning, at-sight attack. The late H. M. Tomlinson describes, in one of his novels, which were as much 'documentary' as fiction, the state of permanent anxiety induced in bush travellers through country inhabited by banteng. Much the same reputation attaches to *Bos* (*bibos*) *gaurus*, the gaur, an Indian and Assam-ese relative of the banteng. In a communication [1] C. R. Stonor says that the mystery is how such a savage animal as the gaur could have changed its temperament early enough in the first stages of domestication to make it worth the people's while to persevere with the work.

The Indian name for the domesticated gaur is *mithan* or *gayal*. It seems impossible to sort out where exactly, or in what context, one name rather than another is used. Just when and where the domestication was accomplished is not possible to say; but the *gayal* is, chiefly, the domestic cattle of the Nagas of Assam to whom it is important, as it is to other tribes of Assam and Further India, as a sacrificial animal. Maybe the domestication was accomplished in that

[1] Published by Zeuner (1963). *See also* Stonor (1950; 1953; 1957).

part of the world rather than in India proper. The most interesting thing about it in our context is the Naga practice of arranging for their *gayal* cows to be mated with wild gaur. Salt-licks are placed in the forest to attract wild cattle; as soon as the *gaur* bulls are in the habit of coming to the licks, *gayal* cows are driven to the spot and left there, so that in due course mating takes place. 'This practice may shed light on the development of some of the *primigenius* breeds in prehistoric Europe. It altogether represents a system of animal husbandry in which something is done to prevent degeneration.'[1]

As for the banteng, nothing whatever is known about their domestication. It may have been accomplished by the Javanese; but when, where and in what circumstances are questions which will be answered only when archaeology has made much greater progress in Indonesia.

[1] Zeuner (1963).

8

Man and the Insect Economists

SILK-MOTHS

Most of the tales current in the Occident touching the origin of silk-moth culture in the western half of the world are legends. I do not mean to say that they are totally untrue, only that they are misleading as they stand. Silk is a commodity much older in the west-Asian/east-European part of the world than is generally realized, and the same is true of the Roman world. Moreover the first silk in Europe was not from China; it was European-made. It is simply the case that Chinese silk and the Chinese silk-moth (*Bombyx mori*), so completely domesticated that it can no longer survive as a species without the care of man, early became so much more important economically than any others that they have come to be thought of as the only originals of their kind.

Since the method of presentation I have tried to use in this book is to work backwards to origins, we should begin with the European story of silk, for it is relatively late. The tale is twofold.

The earliest source of silk yarn in Europe was the island of Cos. Unfortunately very little seems to be known about this ancient silk industry, but since Aristotle (384–322 B.C.), mentions it, it cannot date from later than the fourth century B.C. and may have been older. The fabric produced by this industry was gauzy, semi-transparent, very light, and was at first called in Latin *coa vestis*. It was woven from yarn spun from the cocoon silk of a moth called *Pachypasa otus*.[1] By the

[1] Zeuner (1963).

time of the elder Pliny this rare luxury material was called *seres* (i.e. silk), and its invention attributed to a Coan woman named Pamphile, daughter of Plateas, who 'has the unquestionable distinction of having devised a plan to reduce women's clothing to nakedness'.[1] It was she who discovered how to unravel the threads made by a caterpillar about its cocoon.

An interesting point about this passage in Pliny is that the word *seres* (which had displaced *coa vestis*), from the Greek *serikos*, is, according to the Oxford English Dictionary, probably of Chinese origin. I think this means that before Pliny's time the name, which, with the fabric itself, was reaching Italy from the Far East by way of Persia or the Middle East, was also being applied to the ancient Coan silk; and that this transference has been a source of confusion.

Pliny's notion of how Coan silk was made is confused, acquired at second hand, and then probably misunderstood. What he says comes to this, that on the island of Cos a certain kind of small butterfly, in order to protect itself against the winter cold, makes itself a little jacket by scraping together the down of leaves with its feet, compressing this down into fleece, and coiling the fleece around its body.The creature was exploited by the Coans: they put the butterflies into vessels in a warm place and, using moisture, stripped off their little woollen coats and then thinned the fleece into fine threads. This reads like a badly garbled version of the process of dropping silk-moth cocoons into warm water and winding off the silk loosened by the warm moisture. And there is, of course, no reason why the Coans, just as well as the Chinese or any other people who had a suitable and sufficiently abundant silk-moth in their country, should not have worked out this very simple process. In short, it seems to me

[1] Pliny XI. xxvii.: 'Nor have even men been ashamed to make use of these dresses because of their lightness in summer . . . all the same we so far leave the Assyrian silk-moth to women.' The reference to Assyria is because by this time silk was reaching Italy from the East.

more likely that some of the eastern Mediterranean Greeks did, in fact, during or before the fourth century B.C., discover for themselves how to use cocoon silk and to domesticate the moth *Pachypasa otus* probably to the extent of preserving the cocoons alive (those 'vessels in a warm place'), than that they learnt this craft from the East.

However, it is conceivable that the Coans had somehow learnt the craft of rearing silk-moths from an eastern neighbour who had had it from India or even China, both of which countries had silk very much earlier. But it was Coan silk, not silk from the East, that formed the transparent gauze material which Elagabalus was the first emperor to wear, to the great scandal of contemporary moralists who had no more taste for 'high camp' behaviour than have our own. That was in A.D. 222, so obviously the Coan silk industry lasted for many centuries. But there can never have been very much of it and, like Oriental silk when that began to reach the West, it was extremely dear: even as late as the third century A.D. silk still cost its weight in gold (literally). The real reason why emperor after emperor made sumptuary laws against its use was not a moral but an economic one: the importing of silk was a large contributory cause of the gold-drain from west to east which repeatedly upset the Empire's balance of payments.

Chinese and Indian merchants were so anxious to retain their near-monopoly of silk (Cos can never have been a very serious competitor), and to obtain the highest possible price for it, that they exported only the finished cloth, never the yarn and much less the silk-worms. But as to the *Bombyx* moths, the Chinese imperial policy of marrying princesses of the royal house to powerful foreign, and even barbarian, princes led inevitably to the breaking of the monopoly. Royal ladies had always, as we shall see, taken a hand in rearing silk-worms, and they could not be prevented from taking *Bombyx* eggs with them to their new countries. Meanwhile the Italian customers for silk overcame the Chinese refusal to sell yarn by buying cloth, unweaving it, redyeing the recovered yarn, and then reweaving silk cloth in patterns to suit their own taste. Finally, as the spread of silk-worm rearing to China's neighbours and allies-by-royal-marriage soon led to the breaking of the old monopoly, by the fourth century an Italian or Gaul did not need to be a millionaire in order to afford silk.

Although, as we have seen, it is not true that no silk-moths were reared, or that no silk was made in the West until the smuggling of *Bombyx* eggs into Byzantium during the reign of Justinian (527–65), the story is not entirely without foundation. First, the existence of *Bombyx* silk-moths had been known to a few European *savants* since the fourth century B.C.; Aristotle describes the insect and how the silk was obtained from it. Much later, in A.D. 166, the ambassadors whom Marcus Aurelius sent to the Chinese court saw silk-worms at work. In short, for centuries there had been men in the West who must have wanted to obtain the Chinese silk-moth for their own parts of the world. The yarn of its silk was much stronger, much heavier than that of *Pachypasa otus*, and the silk much warmer. Yet, as far as we know, the *Bombyx* moth, although it had long since become established far beyond the frontiers of China, did not reach any part of the Hellenistic world until the sixth century. In 536 some travelling monks, either Syrian or Persian, brought *Bombyx mori* eggs, smuggled in a hollow cane, to the Empress Theodora in Byzantium. This ex-prostitute was an extremely clever and sensible woman, and it was probably due to her that the eggs were cared for and the caterpillars reared. But it was Justinian who, as soon as he saw the potential profit-ability of the young industry established in this manner, had the good sense to nationalize it at once, thus preventing the industry from spreading westward and spoiling the market, and also retaining fat profits for the State. Of course, his State monopoly could not last for ever, but it did last for some centuries.

Bombyx mori and the other *Bombyx* species, whose caterpillars envelop their chrysalids in a silk cocoon, are all specific
feeders, confined to leaves of the genus *Morus*. The mulberry
which they seem to prefer—at all events it is the one used in
all large-scale sericulture—is *Morus alba*; but the caterpillars
will eat and thrive on leaves of *M. niger*. This was one of the
facts that made it possible for silk-moths of the *Bombyx*
genus to become established in the West when the eggs at
last arrived. *M. alba* was unknown in the western world at
the time of that introduction, and probably did not reach
any part of Europe until late in the Middle Ages.[1] But
because of the value of its deep red fruits *M. niger* had been
introduced from Persia into Anatolia and Greece at least as
early as the fifth century; its fruits were valued not only as
food and for making a wine, but also, owing to their richly
coloured juice, for use as cosmetic (rouge) and even, according to I Maccabees 6: 34, to provoke fighting rage in war-
elephants. At all events, when the silk-worms specific to
mulberry appeared in the West, the mulberry was there
abundantly.

The two Far Eastern lands which constituted the principal
sources of silk were China and India. Thus, the story of the
smuggling of *Bombyx* eggs into Byzantium always gives
China as the country of origin. This is most improbable,
and I think impossible; the time taken to travel from China
to Byzantium in the sixth century was such that the eggs
would have hatched and the young caterpillars would have
died of starvation on the way. In any case, there were by then
much nearer countries which had silk industries, Turkestan
for one, and Persia at no great distance from Byzantium.

---

[1] Heyn (1888) points out (note 73) that whereas Albertus Magnus
describes *M. niger* in *De Vegetabilibus* (6. 143), he does not describe
the white mulberry, though some of his editors have wrongly
claimed that he did. Heyn also refers to Ritter (*Erdkunde*, 17.495),
who found no trace of *M. alba* anywhere in Europe throughout the
Middle Ages. Not even *M. niger* was certainly in, e.g., England
before the mid sixteenth century.

Both these countries certainly had their own *Bombyx* stocks, at several removes, from China. But the Indian case is different: like the Coans later, the Indians domesticated silk-moths of their own, although they may perhaps have had the idea of so doing from China in the first place.

India has about four native genera, about eight species, of silk-making moths: *Bombyx*, five species, living on mulberry; *Anthera mylitta*, living on oak; and species of *Attacus* and *Philosamia*. *Anthera* also has Japanese and Chinese species; these Antheras yield a hard thread which is spun and woven to make tussore or tusser silk. I think it possible that *Anthera mylitta* was domesticated in India before any of the Indian *Bombyx* moths. Tussore or tusser is from Hindi *tasar* which is from Sanskrit *tasara*, a shuttle, and *Anthera mylitta* is the source of tussore silk; so, that insect is the shuttle-moth, was so-called in Sanskrit, and may therefore be the original silk-moth of India. It is true, however, that such terms are very easily displaced and cease to be specific.

Zeuner (1963) says that Sanskrit writings indicate the Brahmaputra Valley as the place, *c.* 1000 B.C. as the time, of the earliest Indian silk industry which later spread into the Ganges Valley. As the cultivation, spinning and weaving of cotton is very much earlier than this in India,[1] it would only have been necessary to adapt the devices and techniques of cotton spinning and weaving in order to suit the new silk thread.

The relative earliness of this date, 1000 B.C., does not of itself imply independent discovery of the use of silk insects and their consequent domestication; the practice of rearing silk-worms could well have reached India from China, by way of Burma, down the Brahmaputra Valley.[2] And it is a fact that the domestication of the silk-moths is far earlier in China than in India—so ancient, indeed, that we know very little about it.

[1] *See* Hyams (1971).
[2] Zeuner (1963).

First, here (for what it is worth) is the legend. At a date equivalent to our 3000 B.C. a hero-emperor, Fu-chi, whose historicity seems doubtful, conceived the idea of using silk of the *Bombyx* cocoon to spin yarn and weave cloth. The implication is that silk was the first woven fabric in Chinese history. But, as Zeuner says, it is virtually certain that before that date the proto-Chinese were weaving wool and some vegetable fibres; and, I would add, they may even by then have had cotton. At all events another emperor, Huang-ti by name, promoted sericulture by charging his empress, the lady Hsi-ling-shi, to teach the people the craft of rearing silk-moths. We get into rather more reliable history with the emperor Yu (*c.* 2220 B.C.), during whose reign thousands of mulberry trees were planted in reclaimed land, and *Bombyx* eggs distributed to the population.

The dates are probably not reliable, but it does seem certain that before 1200 B.C. there existed a ritual cult of the Empress Hsi-ling-shi, which was already very ancient indeed and involved the reigning emperor's consort in the task of officially 'opening' the silk-worm rearing season every year at a religious ceremony, or at all events a ceremony with an ancient ritual. Moreover, so long domesticated has *Bombyx* been, so changed by breeding and selection to produce different kinds and qualities of silk, that the moths can no longer even get out of the greatly enlarged cocoons without human help and have ceased to be viable in natural conditions.

We are therefore driven to the following conclusions. First, a silk-moth of the genus *Bombyx*, probably *B. mori*, was domesticated in China in the third millennium B.C., retained in China as a domestic insect for thousands of years, but later disseminated throughout those parts of the civilized world with climates such that the mulberry trees and the silk-worms could flourish. The Chinese were the first to keep the silk-moths under cover, feeding the caterpillars on racks. Meanwhile, in the second millennium B.C., probably later rather than earlier, another silk-moth, *Anthera mylitta*, was domesticated by an Aryan (Sanskrit-speaking) people in the Brahmaputra Valley; and later other *Bombyx* moths elsewhere in India. Finally, in the fifth or fourth century B.C. the Oriental Greeks of Cos domesticated yet another species of silk-moth, *Pachypasa otus*, the fine silken fabric derived from which was probably the first silk to be seen in Europe, and certainly the first to be made there.

These silk-moths belonging to the *Bombycidae*, the *Saturnidae* or the *Lasiocampidae* are the only lepidoptera ever to be domesticated by man, who, despite the importance of grubs in the diet of all really primitive peoples, has on the whole neglected the exploitation of the insect world. No other insect is economically as important as the silk-moths, even in these days of man-made fibres. But one other insect, *Apis mellifera*, the Honey-bee, a member of the hymenoptera, runs it pretty close and is very much more ancient in domestication. It is perhaps hardly surprising that men and bees get on well together: millions of years before the appearance of *Homo sapiens* the bees had invented social co-operation and division of labour, economic stock-piling and specialization, and carried them much further than man has done. Zeuner, too, has suggested that whereas the bee's self-domestication is and long has been complete, that of man, who must for stability's sake go the same way, is hardly begun. But this sort of comparison is shallow; the bee's technology has been biological, involving modification of their own bodies; ours is mechanical, involving the extension of our bodily powers by means of tools so that, for man, stability with flexibility should be possible.

THE HONEY-BEES

Throughout the prehistory and history of mankind until only about five centuries ago, honey was for most of the human race the only source of sugar; and beeswax was a very important raw material, as it still is. A few peoples had other sources of sugar before the world-wide distribution of sugar canes and sugar beets. They included the East Indians, whose country's flora includes the wild sugar canes; the few Palm-people who had the jaggary sugar yielded by palm sap; and a few tribes of North American Indians who had discovered maple-syrup. For the rest of the world, there was honey or nothing.

The presently existing civilizations all inherited the craft of bee-keeping from their parent cultures. We have improved on the old methods but we have not altered them, for the bee-keeper's business is simply to accommodate as best he can bee-communities whose laws are far more rigidly un-changeable (and whose needs are, therefore, always the same) than 'the laws of the Medes and Persians which alter not'. We know much more than did our fathers about bees and their social organization—for example, there is the discovery by Roesch and van Fritch that they communicate very exact information to each other by means of their dance-language —but the use we make of this knowledge has not yet, in practice, made much difference to bee-keeping.

The bee-keeping we practise in the West we received from the Romans, whose farmers already knew that bees have a complex social life which has to be respected if we are to exploit the insects; knew how best to site an apiary, and had worked out good practical shapes for hives; knew what flowers to plant in order to provide the bees with the best sources of raw material for their industry; knew how to feed the colonies in the winter-time, and understood the advantages and means of providing weak colonies with new queens.[1]

But the Romans had not, any more than we have, had to discover and invent all this for themselves. They had it from much older cultures. The Etruscans must have taught them a good deal in this as in other respects; and certainly the Greeks and Carthaginians did, although perhaps it is fair to

[1] Varro in *De re rustica* (37 B.C.) and the Fourth Book of Virgil's *Georgics*.

say that the Romans found out for themselves a little more than Aristotle had known about bees in 343 B.C., the approximate date of his *History of Animals*.[1] Yet the technically advanced Greek bee-keeping of the sixth and fifth centuries B.C. was a fairly recent phenomenon, for Homer (eighth century) knows nothing of domesticated bees at all, only of gathering the honey of wild bees. So that some time in the seventh century the Greeks learnt bee-keeping from somebody. From whom? Probably from the Cretans: was not Zeus reared by the daughters of King Melissus on milk and honey? And did not the honey come from the hive-bees of the princess Melissa? Melissus and Melissa mean honey.

There is no reason, however, to think that the Cretans themselves conceived the domestication of the honey-bee, although they may, once they had the idea, have hived wild swarms. It is more likely that they received the art from Egyptian mentors; or they might have learnt it from the Philistines and Canaanites, who by their agricultural skills had made Palestine into that 'land flowing with milk and honey' which was promised to the Hebrews by their tribal god. It was from Tyre and Sidon, of course, that their Carthaginian colonists took bee-keeping to North Africa, where their brand of beeswax—*cera punica*—was noted as the best in commerce.

The pre-Hebrew peoples of Palestine seem to have been keeping domesticated bees before the more ancient cultures to the north in Mesopotamia;[2] yet it was not from the Canaanites or Philistines that the Mesopotamians, Babylonians and Assyrians (for the earlier Mesopotamian cultures did not know honey at all) derived their bee-keeping. A cuneiform text dated 1100 B.C. reads:

I, Shamash-resh-ussur, Governor of Sukhi and Ma'er, have brought from the mountains of the people of Khabkha into the land Sukhi the bees which collect honey and which

[1] Books 1–8. Book 9 is more advanced but it is by the Pseudo-Aristotle, is later, and written by a professional apiarist.
[2] The Israelites conquered Canaan *c.* 1200 B.C.

since the days of my ancestors nobody has seen. I settled them in the gardens of the city of Gabbarini, so that they should collect honey and wax. The manufacture of honey and wax I understand and the gardeners understand it also. Whosoever wishes to raise his voice, let him ask the elders of his country, 'Is it the truth that Shamash-resh-ussur, Governor of Sukhi, introduced honey bees into the land of Sukhi?' [1]

In most of the cases I have dealt with in this book and in *Plants in the Service of Man*, it has been necessary to award, as it were, more 'firsts' to the proto-civilizations of Mesopotamia than to Egypt. But it is quite clear that, in the case domestication of the honey-bee, Egypt comes first. Before 2600 B.C. Egyptian bee-keepers had discovered how to keep both the honey and the brood. It is probable that even after Neolithic man had become a bee-keeper, instead of a mere bee-robber, he still had to kill or drive off the bees before he could get at the honey; and to rely for a new colony on a wild swarm for which he would provide a hive. The Egyptians used smoke, as we do, to paralyse or keep down the bees while they took the comb from the hive. It is also probable that they were the first bee-keepers to leave the bees a provision of honey, and so keep a colony in being; perhaps, too, they were the first, by piling up pyramids of hives (*see* below) to give the bees so much room in their immediate neighbourhood that colonies were not lost by swarming. In short, by about 3000 B.C. Egyptian bee-keepers had discovered the economic value of that state of permanent anxiety which the Israelite Joseph, as Pharaoh's chancellor of the exchequer, was to translate later into human terms and which drives bees, as it drives bankers, to wear themselves out in excesses of hoarding.

The most primitive hive used by bee-keepers was a hollow log hung vertically or horizontally. But the Egyptians had few trees and were soon using earthenware cylinders instead,

---

[1] Quoted by Zeuner (1963).

plugged at one end, open at the other. This form had, of course, been suggested by the hollow-tree hives of wild bees.

It is impossible to assert that the honey-bee was first domesticated in Neolithic Egypt; the story goes well beyond 3000 B.C., and the most one can say is that this domestication occurred in the Neolithic period and that it was probably undertaken quite separately in a number of places; in fact it must have been, for, to take an extreme case, when Cortés reached Mexico in the early sixteenth century he found honey to be an important article of diet among the Aztecs of Tenoch-titlan, and there is not a trace of evidence that the Aztecs or their still more civilized predecessors in Central America could have learnt bee-keeping from the Old World.[1]

Having worked backward to the date 3000 B.C., it will be convenient to start again, at the beginning this time, and work forward to it.

Nobody can ever know at what date man began the practice of robbing the hives of wild bees. *Apis* is not the only genus to produce a species of honey-bees: there are also species of *Melipoma* and *Trigona*, and the distribution of honey-bees of one sort or another is very wide indeed, as one might expect of an Order which, having a history of at least thirty million years, is older than the present geographical shape of the world. It is not, in this case, simply a matter of saying that the practice of robbing the hives of wild bees began with Palaeolithic man; even bears, after all, rob hives, and for my part I am sure that so did those species of the genus *Homo* which preceded our own on earth, in other words that bee-robbing goes back to *Homo neanderthalensis*—perhaps even to *Pithecanthropus pekinensis*. Robbing hives of honey is, of course, a long way from domesticating bees, but it must have suggested it.

---

[1] Some early Spanish authorities say that an example of Aztec *haute cuisine* was a *mousse au chocolat* made by beating chocolate into honey.

Palaeolithic man was a great artist, as we know from his works at Lascaux and elsewhere. But he left no picture which can give us an idea of how he dealt with the problem of stealing honey; the earliest picture, a rock-painting, is possibly Mesolithic, probably early Neolithic, and I shall come to it presently. However, since it is always painful, often dangerous, and occasionally fatal to rob a hive of wild bees, it is probable that smoke was used, as we still use it, once man had control of fire, perhaps even before he appeared as *Homo sapiens*. Bush fires would have taught the observant hunter how wild honey-bees behaved in case of fire, and man was as capable 30,000 years ago of learning from observation as he is today. There is another possibility: observation of bears robbing hives would have suggested an armour of bear-skin or other animal hide.

The painting I have mentioned above, probably Neolithic but possibly Mesolithic, is in the Cueva de la Araña in eastern Spain. It depicts the following scene. The background is a cliff-face, high up on which a man (or, judging by the width of the hips, a woman) is putting one hand into a hole or niche in the stone, while a number of bees fly around, perhaps attacking. True, they are much too big to be bees, but one does not expect strict realism from the rock-art of the Neolithic epoch. The man (or woman) is standing on a rope ladder fixed to an unidentifiable support represented by two strong horizontal lines above the hive niche. Another human figure stands on the rope ladder far below; the person at the hive has a bag in his (or her) right hand.

There can be no question but that this curiously vivid little picture represents a human being robbing a hive of honey. But there has been much controversy about the nature of that element of the picture which I have called a rope ladder and which to my mind quite obviously *is* a rope ladder. The three vertical lines have been identified as various things, even, quite implausibly, as some kind of tree. Zeuner (1963) asks why, if they are meant to be ropes, there should be three where one would suffice. The answer, obvious to anyone who has done any clambering of that kind, is that one rope would not suffice: the honey-gatherer had to have one hand free to reach for the comb, and he had to manage the bag. Nor do I understand why the horizontal lines, quite thick and strong in one case, connecting the three ropes, have been overlooked, and, even more strange, the physical attitude of the man lower down the ladder.

This picture should, I believe, be interpreted as follows. The man at the niche is kneeling on a cross member of the rope ladder and has his left elbow hooked under another to steady him while he collects the comb with his left hand. This attitude would be impossible were it not for the man on the ladder below, whose weight holds the upper part of the ladder steady. That the controversial element is indeed a rope ladder, is apparent to anyone who has ever climbed a three-rope ladder, from the attitude of the lower figure: his weight, being on his hands, is holding the ladder rigid for his companion above; inevitably his body curves under, and his feet push the lower part of the ladder outwards. Two problems remain: first, to what is the ladder tied? One imagines several answers, but there really is not sufficient detail to enable one to prefer one to another. Second, how does the collector deal with the bees which seem to be attacking him? From the slender shape of the figure, he or she is not clad in skins. Either he puts up with being stung, or he is smeared with a bee-repellent, some kind of animal or vegetable unguent. This latter is not an outrageous sug- gestion: had prehistoric man been less ingenious than we are, we should not be here, ensconced in the nuclear power age; as capable of clever invention as we are, he was handi- capped only by having a much smaller corpus of science behind him.

But all this refers to hive-robbing. How was the step made from that to bee-keeping? Very easily indeed: people used to hive-robbing would have become acquainted with the behaviour of bees, including the act of swarming to establish a new colony. All they had to do was to provide a suitable hollow log or other potential hive in which the bees might settle.[1] The bees themselves may have shown men how to capture them by moving into a vessel—pot or basket—standing empty in a camp or settlement.[2] In India, the Canary Islands and probably elsewhere where there are still troglodytic communities, bees sometimes hive up in caves already occupied by man. At all events, man was certainly keeping bees in crude hives made of hollow logs before the end of the Neolithic, because, to return now to Egypt, the 'bee was an important member of the economic-religious pantheon from the First Dynasty onwards'.

The silk-moths and the honey-bees, then, are the only two insects domesticated by man which are of great economic importance. But there is one other insect which was more or less domesticated, once had economic importance enough to make great fortunes for its exploiters, and is today again rising in importance owing to the failure of some synthetic dyes on certain man-made fibres. It is of peculiar interest that a 'natural' dye made from the bodies of insects, having been superseded as a result of progress in industrial chemistry, is now enjoying a revival because of further progress in the same applied science.

THE COCHINEAL INSECTS

It is, perhaps, cheating a little to write about scale insects in the context of domestication. No attempt has, so far as I know, ever been made to modify or improve them (from man's point of view) by selection or breeding. Yet if you systematically plant out the host plant of an insect in order

[1] In this connection see Zeuner (1963).
[2] See Fraser (1951).

to feed that insect and cause it to multiply, I do not think that the word 'domestication' is altogether out of place.

Historically, the most ancient in human service of these minute insects is, so far as we know, the lac insect. I say 'so far as we know', because it is conceivable that the Central American Indians—Maya, Toltec or Aztec—were using cochineal even earlier than the Indians of India were using lac; but, given the dates, it is extremely unlikely.

Laccifer lacca is a scale insect which lives on twigs and young branches of *Acacia*, *Ficus* and some other genera in India, Burma and south-east Asia. The creature is equipped with a relatively long beak, which it plunges into the bark and is thus able to suck the sap. While doing this it excretes or secretes a resinous matter which hardens into a mass full of holes. The female insects become embedded in this and never escape from it, because following fertilization their bodies swell enormously, becoming a crimson sack of young, after hatching which, they die. The process of exploiting them is very roughly as follows. The twigs, loaded with their coat of insects and resin, are gathered twice a year—in June and November. The red lac dye is extracted by immersion in a solvent—primitively, even in hot water. The lac is then melted off the twigs to produce the shellac of commerce, which, according to the species of host-tree in use, the stage reached in the life-cycle of the insects and perhaps the strain of insects, may be orange, brown, nearly black or—when chemically bleached—completely transparent.

The East India Company was responsible for introducing lac dye and shellac varnish into Europe, in the late seventeenth century, and it has been a valuable article of commerce ever since. But for how long had it been in use in India by then? For at least 3,000 years. Credit for the 'domestication' of an insect whose name (*lac*, *lakh*) means 'one hundred thousand' in reference to the quantity required to make a useful amount of lac or shellac, belongs to the Aryan invaders of India who were using lac not later than 1200 B.C.

The Greeks were the next people to make use of a scale insect. Whether they had the idea from the East or thought of it for themselves is not clear. The insect in question was *Lecaninum ilia*, and the product the crimson dye called *kermes* of which the Romans also made use.

The third of these scale insects to be 'domesticated' was a Mexican one, *Dactylopius coccus* belonging to the family Coccidae, which lives chiefly on the *Opuntia* cacti, especially *O. coccinellifera*. The Spaniards, when they arrived, found it already in 'cultivation' or 'domestication' by the Aztecs, and it must by then have been long in use because it, and its cactus, had been introduced into Peru. Like the lac insect, the cochineal insect is minute (about 70,000 individuals weigh one pound). The dye which they yield is either crimson, scarlet or orange. The insects, brushed off the cactus plants with a light brush into a bag, are killed by heat applied in one of several ways—different methods are used in Mexico, Honduras, the Canary Islands and Algeria—and processed to extract the dye coccinealin. In the past, large fortunes were made from this industry in the Canary Islands. Then came aniline dyes, and the bottom dropped out of the cochineal market as it did out of the indigo market. But not only is cochineal still useful for the safe dyeing of foodstuffs and in making cosmetics; it is again being used as a dye on some man-made fibres which will not take synthetic red dyes very well.

Other uses have been made of insects. One thinks of the training of circus fleas; of the Chinese singing crickets and fighting crickets, and wonders why no man has ever tried to breed and select super-fleas and super-crickets. Perhaps it was scarcely worth while. But that argument will hardly do in the case of the *Cochliomya* fly maggots which live only in and on the morbid flesh of infected wounds. Centuries ago the great Ambroise Paré discovered that war wounds infested with maggots healed better and more quickly than wounds free from maggots. His discovery was repeated by

W. S. Baer in 1916 and he effected remarkable cures of long-standing ulcers and bone infection by infesting them with maggots. Doubtless, had minds been set to it, we should long since have 'domesticated' an antibiotic *Cochliomya* whose application would at least have been free from side effects.

9

The Reindeer

At the present time the several species—or perhaps they are
no more than geographical races—of reindeer are confined
in their range to the extreme north of Eurasia and America.
Their scientific name, *Rangifer tarandus tarandus,* covers a
group of deer, including caribou, closely related but with
minor differences, which inhabit Alaska and northern Canada,
Greenland, northern Scandinavia and Siberia. In those
countries the reindeer are still of considerable economic
importance, providing some tribes with their only means of
livelihood. But formerly, when the whole northern hemisphere
down to the latitude of the Mediterranean was about as cold
as our extreme north now is, and its topography and flora
were sub-arctic, the natural range of these animals was
enormously more extensive.

Before the last, the so-called Würmian, ice-age, the climate,
flora and fauna of Europe and North America were sub-
tropical. But with the advance of the glaciers and the steady
fall in temperatures all over those regions, the flora and fauna
had to change. Such animals as elephant and hippopotamus
were confined to much lower latitudes, and the vast area
which they vacated was occupied by the arctic fauna, beautiful animals like white fox, snow-hare, snow-leopard (in the
East); and reindeer. This last was to become the most important species—as a source of meat and industrial raw materials
—in the economy of men who were in the Upper Palaeolithic
phase of culture; [1] so important, indeed, that before the
general adoption of the system of naming epochs in our

[1] Arambourg (1968) in Varagnac.

prehistory after types of industry, a whole long age in that prehistory was called 'the Reindeer Age'. Daily life and industry depended entirely on the reindeer. So, even, did the arts: one of the most superb groups of Magdalenian works of abstract art, the carved rods discovered by Mme de Saint-Perier in the Haute-Garonne, is wholly of reindeer horn; a magnificent engraving of a reindeer is the glory of the Combarelles cave at Les Eyzies in the Dordogne valley; and Solutrean burials have yielded a great many fine figurines of reindeer in various materials. It is perhaps no mere coincidence that the brilliant Magdalenian culture moved northward with the reindeer retreating as the climate warmed up again and the glaciers began to shrink, so that they appeared simultaneously on the shores of the Baltic, where later, as eland replaced reindeer moving still farther north, the quality of the arts began to decline.[1]

Since that epoch the geographical, or perhaps one should say ecological, races of reindeer have become more specialized than they formerly were. It is not necessary to go into detail about this, but today there are specifically forest reindeer and tundra reindeer, with different ways of life to suit the different environments. But it appears that during that Upper Palaeolithic phase of European culture no such specialization had yet occurred, or at all events it was much less marked; and of the fossil reindeer remains found throughout Europe, including Russia, Zeuner (1963) says that they seem for the most part to be something between our own forest and tundra types.

The same great authority is fascinating on the subject of the reindeer's peculiar eating habits. Although reindeer feed chiefly on ground lichens and the lichens which, in the far north near the limit of the woodland belt, are epiphytic on trees, the deer have a very marked taste for woodland fungi. This is important in our context; for reindeer eat the poisonous species of fungi as well as the wholesome species; the

[1] Breuil (1968).

toxin thus absorbed makes them sluggish and sleepy, and in that state they are very easily captured, a fact of which pre-historic man doubtless took full advantage. Also important is the fact that reindeer readily adopt a carnivorous diet: they eat both meat and fish, and so might well be tempted to approach the edges of human encampments for discarded bits from the hunter's bag or the fisherman's catch, rather in the way that wolves, which were to become dogs, were attracted to man. Still more remarkable, reindeer will hunt and eat lemmings.

They have a craze for salt. Besides eating all kinds of sea-weed, they drink not only sea-water but also urine, especially human urine and dog's urine, both of which they prefer to their own, though they drink that also. Many and various explanations have been advanced to account for this curious taste, none of them quite satisfying. Perhaps it is just the salt in the urine that they like. I have read somewhere of one of the extreme northern peoples—I no longer recall which and do not have the reference—who are in the habit of drinking their own urine after eating intoxicating mushrooms, thereby enhancing the toxic effects. Perhaps it is too much to suppose that reindeer, with their strange taste for toxic fungi, use their kidneys as a 'still' in much the same way.

What concerns us is that the reindeer's taste for meat and fish, salt and urine must have made it much easier for man to gain a hold over them, with so many baits to offer. So also must another of their attributes as a species:

The reindeer is the most gregarious of all the Cervidae as, except for the old males, it is most unusual to find single individuals. The herds which form are small in the forests; whilst in the caribou they vary from a few individuals to a few hundreds. There may be more occasionally, but really large herds are observed in the tundra. A hundred and seventy years ago Pallas described how, in north-east Siberia, he first saw a few herds of two or three hundred head each, which were followed by thousands and hundreds of thou-sands, making one mass of animals ten miles or more across.

Their antlers appeared like a moving forest. When crossing
the Anadir the animals crowded so closely together that they
could not avoid being killed by the natives because they could
not even step aside, and the fawns were using the backs of
adults as a bridge. There are many other reports of this kind
and there is no reason to think that the numbers congregating
on migration have lessened since. The Magdalenian carving
from the cave of La Mairie, Dordogne, provides a vivid illus-
tration of some such herd which occurred 20,000 years ago.[1]

There is yet another characteristic which would have
helped to put the reindeer at the mercy of primitive hunters:
the regularity of the movements of those great herds. In
summer they shifted northwards, away from the forests, in
winter south again. The reason for those movements was
summer heat, too much for the deer which are equipped by
nature for extreme cold; Zeuner has suggested that they may
have moved with the light, that is moved north as the days
lengthened, south as they shortened. But the point here is that
year after year they followed the same route, so that the
Palaeolithic hunter could count on finding the reindeer always
at the same place at the same time—for example at a particular
ford of a particular river at a given time of year.

[1] Zeuner (1963), who quotes the following from the *Ottawa
Naturalist*, 1917. The Magdalenian engraving in question shows
reindeer at the head and tail of a forest of moving antlers.
  'He stood on a hill in the middle of the passing throng with a clear
view of ten miles each way and it was one army of caribou. How
much further they spread he did not know. Sometimes they were
bunched, so that a hundred were on a space of one hundred feet
square; but often there would be spaces equally large without any.
They counted at least one hundred caribou to the acre; and they
passed him at the rate of about three miles an hour. He did not
know how long they were in passing this point; but at another place
they were four days and travelled day and night. The whole world
seemed a moving mass of caribou. He got the impression at last
that they were standing still and he was on a rocky hill that was
rapidly running through their host.
  'By halving [these figures] to keep on the safe side, we find that
the number of caribou in this army was over 25,000,000. Yet it is
possible that there are several such armies.'

Despite the very close association between the reindeer and Palaeolithic man in Europe, it is not suggested that the animal was in any sense domesticated by him. No doubt some communities of men lived parasitically on a herd by moving with it; but although that is a kind of association between man and beast, and as such is a step towards domestication, it is far short of being domestication, for it is the beast which calls the tune, not the man. It is virtually certain that the hunters did follow moving herds, just as packs of wolves did, killing as they went; but following a herd for a few days' journey from your lair, whether you be wolf or man, is very different from becoming so completely nomadic that your movements as a family or tribe or clan depend entirely on the herd's movements which you slowly learn to control.

The wolf is not introduced here merely by way of making a comparison. Zeuner (1963) suggests that since wolf and man had somewhat similar practices in following the reindeer herds and killing as they went, and since by Mesolithic times the wolf had been taken into domestication and had become the proto-dog, the domestication of the reindeer may have been connected with that adoption of the dog. Certainly the co-operation of man and wolf would have made control of small herds of deer—the first step towards domestication— much easier than it would ever have been for man alone, just as it is easier for a shepherd to control a flock of sheep with than without the help of dogs.

The domestication of the reindeer, then, began when men, helped by dogs, established a measure of control over the movements of a herd, although at first this may have amounted to no more than heading it off into some kind of trap or ambush. The scene is not hard to imagine: the selection of a natural cul-de-sac or the construction, with stakes and cut thorn bushes, of an artificial one; the skin-clad hunters, with their throwing sticks and flint knives, running in crescent formation behind the packed herd of deer, driving them with shouts and yells, while the wolf-dogs, racing ahead, by snapping and snarling turned the leaders in the direction their partners wished the herd to take. The *dies irae* for the grazing and browsing animals of the world was that day when a boy took a wild wolf cub to be his playmate, and the long partnership between man and dog began. But because what began as bane for deer and cattle and sheep, as a partnership in hunting and killing, suggested in due course the possibility of the complete and permanent control of herds, and ended in domestication; and because that domestication has preserved and diversified and improved whole species almost beyond belief, that day of destruction was also a day of creation.

There were two widely separated stages in the domestication of the reindeer. In the first, man, helped by his partner the wolf-dog, gained control of herds simply in order to use them as meat and raw materials on the hoof, living on those herds nomadically as other peoples lived on herds of cattle or horses. In the second stage, the reindeer nomads learnt from the examples of more advanced peoples who by then had their herds of cattle, horses, sheep or goats, to drive deer in harness, to ride them and to milk the does.

There is virtually no direct evidence of any kind which might enable us to fix even approximately a date for the earliest first stage of reindeer domestication. One can, however, form some idea of how it was done by analogue, from historical accounts of the semi-domestication of llama herds on the *altiplano* of the Andes by the Incas and other mountain peoples of South America before the Imperial Epoch in ancient Peru. A proportion of the animals was brought under complete control; a much larger proportion was guarded and watched, but left wild, to be rounded up at regular intervals in a great *battu* for which thousands of men were recruited; culled, decimated and sheared, the animals were released again to get their own living.[1] Probably some-

[1] See, e.g. Karsten (1949).

thing very like that happened in the case of the reindeer in Mesolithic Europe.

One writer on the subject does, in fact, put forward this view. He found evidence in north-west Germany that at a date which can be fixed at about 12,000 B.C. (it seems startlingly early) most of the local reindeer population was herded, not merely hunted. If this be so, then far from being belated as most authorities claim, the domestication of the reindeer is among the earliest.[1] Sledges which were probably hauled by men but could have been drawn by a draught animal, though too heavy for dogs, are known from the Mesolithic.[2] It has been suggested that they may have been drawn by reindeer.[3] So high a level of domestication at a date which must be around 7000 B.C. seems improbable.

It really is hard to accept these dates; to do so is almost to admit that the reindeer was the first draught animal, and that peoples whose geographical location and environmental difficulties made them relatively backward taught the world to drive animals in harness. Moreover, excavation of Mesolithic sites in the Crimea, a comparatively advanced region, reveals plenty of remains of hunted reindeer, among those of other animals, assocated with microlithic tools, but nothing to hint at domestication. Needless to say, the same is true of late Palaeolithic sites of the bone-carvers of Siberian Mal'ta and of Afontava Gora Magdalenian. The dog appears as a domesticated animal among reindeer-hunters in Siberia, in a sort of retarded and decadent Magdalenian context; but it must have been long after man had the dog to help him that he began to herd the deer.

We can summarize the probable course of events in man's association with the reindeer as follows.

There was nothing in the least resembling domestication of the reindeer during the period when it was common in

[1] Polhausen (1949 and 1953).
[2] e.g. from Saare Jarin, in Finland, and elsewhere.
[3] Sirelius (1916).

vast herds all over Europe, that is to say during the period
which coincides with the brilliant Palaeolithic Magdalenian
culture. In northern Europe and/or Siberia and perhaps
North America, in a belated Mesolithic context, men aided
by wolf-dogs gained a measure of control over reindeer
herds and in due course became nomadic in order to suit the
seasonal needs of their herds. The practices of castrating
male fawns by biting, of using urine and salt-licks to dominate
the deer, of ear-nicking by way of branding, and of using the
lasso to catch and throw the deer, were all developed; [1] but
the only use made of the deer was to slaughter them for meat,
hides, horns, fat and bones.

Later, probably much later, in a Neolithic context, reindeer nomads came into touch with relatively advanced farming communities who had milch cattle and draught cattle,
and even riding cattle or horses. This gave them the idea of
milking, driving and riding their reindeer. Then, and only
then, was domestication of the reindeer complete. Zeuner
(1963) cites the case of the frozen horses found in the Pazyryk
kurgans, in the Altai, dating from the first millennium
B.C., as indirect evidence of reindeer riding: for one of the
horses wears a reindeer mask complete with antlers; this,
he says, seems to convert it into a superior kind of riding
reindeer.

The reindeer nomads who learnt to ride their animals were
those who came into touch with the horse-nomads of the
Altai—Turks and Mongols; for their reindeer saddles are
based on Turkish and Mongol designs. The peoples in
question are the Tungus, Yukagin, Soyot and Karagas; and
the Samoyed, Ostyak and Vogul peoples to the north.[2] Some
of these tribes also use reindeer as pack-animals, again with a
Turkish- or Mongol-type pack-saddle. So also do some
peoples, the non-riding Lapps for example; but they must
have had the idea from quite another source, because their

---

[1] Hatt (1919).
[2] Elisseeff, in Varagnac (1968); Zeuner (1963).

pack-saddles are of a different pattern.[1] Yet other peoples, such as the Koryak and Chukchi, and some of the riding peoples also, use reindeer to draw sledges, an idea which developed out of the Samoyed practice of using dogs for the same purpose.

When the reindeer-herding Lapps came into contact with stock-raising Scandinavians, who by then had herds of milch cows, they learnt to milk their reindeer does. Yet other reindeer nomads, meeting with the mare-milking nomads of north-east Asia, imitated them likewise.

Despite the lateness of these final stages in the complete domestication of the reindeer, the animal would seem to be one of man's oldest servants. Nor is its day yet done, since it can live and thrive where no other domestic cattle can be kept. Thus a new Eskimo-Reindeer economy in North Canada, started by driving a great herd of reindeer across the country to a tribe whose ancient fisheries and fish-economy were failing them, is not yet twenty years old. We are beginning to learn that to discard an old kind of animal-husbandry because we have newer and more high-yielding kinds is, in certain places and conditions, to throw out the baby with the bath water. We cannot afford to dispense with any animal or plant which our ancestors domesticated; on the contrary, we should be doing as they did with species never yet attempted.

[1] Hatt (1919)

10

Old and New World Camels

As a rule zoologists distinguish two species of camels: *Camelus dromedarius*, the one-humped (Arabian) camel or dromedary; and *Camelus bactrianus*, the two-humped (Bactrian) camel. But it seems that this division into two species is of rather doubtful validity: anatomists point out that the differences between the skeletons of the two kinds are insignificant; the leg bones of the Bactrian camel are a little shorter than those of the dromedary. These differences are in fact so small that palaeozoologists, examining fossil camel bones from areas where the two kinds overlap, are unable to decide whether they belong to a two-humped or to a dromedary. Even the differences in the number of humps is more apparent than real, for dromedaries have a vestigial second hump which is present even in the embryo. Moreover Bactrian camels and dromedaries interbreed freely if given the chance (as for example in Turkey), and produce one-humped, vigorous, fertile offspring.[1] It seems likely that *C. dromedarius* and *C. bactrianus* are not two different species, but geographical races of a single species. For our purpose it does not matter much, although the histories of the two kinds have been very different. In any case, these two kinds of camel have been in existence since a very remote past, both as wild animals; it is not a case of one or other kind having developed, as a result of artificial selection, in domestication.

These camels are descended from an ancient parent stock

[1] Curasson (1947).

which evolved not, as one might have expected, in Asia where they have been native throughout the history of mankind, but in North America throughout the Tertiary epoch. In that respect they resemble the horse, as they do also inasmuch as their most remote recognizable ancestors were no bigger than hares. Our genus, *Camelus*, also first emerged in North America, including Alaska but well down into Central America too, during the Pleistocene. Late in that epoch what is now the Behring Strait was still dry land connecting the Old and New Worlds by a land-bridge which was used later by man and his dog to cross into and populate the Americas. By that same land-bridge the camels migrated into Asia, and slowly spread across the dry belt as far as South Russia and Rumania, where their westward movement was checked by the more humid climate to which they were not well adapted. Meanwhile the true camels slowly became extinct in their original native land, although an earlier branch of the family survived, as we shall see, in South America.

When and where the original immigrant stock separated into the two kinds or species, the two-humped camel and the dromedary, we do not know. Zeuner (1963) thinks it was somewhere in western Asia. At all events, while the two-humped camels remained to populate the whole main extent of Asia and possibly the eastern fringe of Europe, within the dry belt, the dromedaries moved into Palestine, Arabia and North Africa, where, as wild animals, they were contemporary with the Palaeolithic Neanderthaler known as Levalloisian Man.

No living man has ever seen a wild dromedary; but they have not been extinct for very long, perhaps for a couple of thousand years, perhaps rather less than that. Why they became extinct is not clear; it is suggested, however, that the desiccation of their habitat was to blame. Fortunately, they had been taken into domestication before that disaster, and a fine species was saved.

The Bactrian camel is almost certainly not extinct in the

wild, only very rare and very hard to find; but it is difficult to be quite sure of this, for camels have been very little modified by domestication. Such typical modifications as reduction in size by comparison with the wild animal in the early stages of domestication, changes in coloration such as piebaldness, shortening of the facial in relation to the cranial part of the skull in the domestic as compared with the wild animal, reduction in brain size and changes in the number of caudal vertebrae—these have either not occurred or have been so slight as to be useless in trying to distinguish between the wild and the tame animal.[1] Zeuner thinks that this may be so because camels are never stabled or penned but continue, under human control, to live very much their normal life. The point is that it is extremely difficult even for a skilled zoologist to tell whether a herd of camels encountered living a natural life in the wild are really of wild stock, or are simply feral (or the descendants of feral animals), that is, once-domestic camels which escaped into the wild. There are, however, two modifications which do occur and may help: Berry (1969) says that 'the development of local fat accumulations (as in the hump of camels) has usually been favoured under conditions of domestication', so that a noticeably smaller hump may perhaps be evidence for genuine wildness, as also, apparently, may a smoother coat.

During the past century two-humped camels living wild, possibly of genuinely wild stock, but maybe of feral stock, have been observed on a number of occasions: by the great zoologist Przewalski in the Lop Nor; by Sven Hedin in the Tarim Basin; by various Mongolian zoologists, who have photographed them; and most recently (1956) by Ivor Montagu in the Gobi Desert.

Wild or feral, such herds are very rare and small. As in the case of the dromedary, the Bactrian camel's numbers have been enormously reduced, to the point of extinction, by the progressive desiccation of their ancient habitats.

[1] Berry (1969). Zeuner (1963), on whom Berry relies.

From palaeontological and other evidence it is clear that dromedaries became confined, in the wild, to the Arabian peninsula and the skirts of the Sahara. They were not quite unknown to the ancient pre-Dynastic Egyptians, but they were not at any time native to the Nile Valley; the very few graphic or plastic representations of camels must have been based on beasts killed, captured or sketched by travellers, for as a domesticated animal the camel did not reach Egypt until very much later. Nor is there any trace of the wild dromedary in Mesopotamia or Persia. It therefore looks as if the only places in the world where the dromedary could have been domesticated are Arabia and perhaps Palestine; we shall return to that consideration presently. On the other hand the Bactrian two-humped camel ranged from west of the Iranian plateau clean across Asia to north China, where, apparently as in Central Asia, there may be some small surviving wild herds.

Camels are not creatures of the agriculturally based urban civilizations; like the yaks and the reindeer, though in a very different way, they are animals which have served man only because of their adaptation to extremely harsh physical conditions. And since the civilized, farming, urban peoples have made very little use of them in historical times, and made none in prehistoric times, I shall not, in this chapter, try to work their history backwards. Long after they were familiar to the peoples of the early civilizations of Mesopotamia as pack-animals used by Arab and other desert traders, and by the Bedouin as war-chargers, they were still rejected by those peoples as domestic animals. There is no mystery about that; from the point of view of people living in crowded cities and practising intensive irrigation agriculture on a restricted area of land, there was no point in keeping camels, which would have been a mere nuisance.

For people who have rich pasture, or who grow fodder for stalled beasts and have other stock, camels are a bad investment; they do not reach breeding age until they are five years old, and then breed only slowly, as do the rest of the Camelidae. In climates with enough humidity for crops to flourish and cattle to thrive, camels are susceptible to a number of troublesome diseases and apt to be more nuisance than they are worth. Neither their meat nor their milk is very good; a fine cloth can be made from their hair, but with much more trouble than from wool. Camels are uncommonly stupid and therefore very difficult and troublesome to train; they are also bad-tempered, the males being notoriously aggressive and often dangerous. They have a very unpleasant smell; man may be prepared to tolerate this, for, as we know,

> They 'aven't got no noses,
> The fallen sons of Eve.
> Even the smell of roses
> Ain't what they supposes . . .

But animals are not so tolerant or so insensitive; unless trained to it, horses and other domestic animals are deeply disturbed and may even be panicked by the stench of camels. While it is true that camels can thrive on poor pasture, they need an enormous range of it in order to do so; which was exactly what the ancient intensive agriculture civilizations did not have. The alternative is to stall-feed them, as in the Canary Islands; but they consume a vast amount of fodder.

In short, from the point of view of the peoples of the early Mesopotamian civilizations there was everything to be said against, nothing for, the camel; and it was not until progressive desiccation had radically altered the environment, that the camels could come into their own, except in the arid lands surrounding the great oasis civilizations.

In those arid lands the situation was very different. There, the quasi-impossibility of supporting other stock, the impossibility of arable farming, the vast regions of drifting sand in which only a camel can walk easily, made the camel a most valuable animal. Consequently it was by peoples of the Arabian and Asiatic steppe that camels were domesticated.

The earliest written references to domesticated camels are not very ancient, for they are Biblical; and the earliest of these is misleading. When Abraham visited Egypt and was received by the Pharaoh, *c.* 1800 B.C., camels were, according to Genesis 14, among the gifts which the Pharaoh gave him, presumably in payment for Sarai's favours; Zeuner (1963) points out that they cannot have been, since Egypt at that time had no camels. He adds that doubtless the Jewish author of this book, unable (as a good Bedouin, I would add) to believe that a royal gift would not include camels, added them of his own accord. Still, evidently domesticated dromedaries were familiar in Palestine by about 1800 B.C. if not rather earlier, because there is also in Genesis (24, 25) the account of Abraham's steward going with ten camels to Nahar in Mesopotamia, in search of a bride for Isaac, and that pretty scene of Rebekah watering Nahar's camels at the well. Archaeological finds confirm this conclusion.

It does not, however, follow that camels were domesticated in Palestine. True, the 'Arabian' part of the country was pastoral and inhabited by nomads; but the skill and industry of the Canaanites, the people who built Tyre and Sidon and Byblos, had turned a good deal of it into a 'land flowing with milk and honey', that is into rich agricultural land, the very kind where the camel had no place. It is much more likely that the domesticated camel was introduced into Palestine from Arabia proper. It was from Arabia that the Midianites, whose camels were 'without number' (Judges 6–8, poured into Palestine and warred with the Israelites about 1100 B.C. And although archaeological evidence found in Arabia—rock engravings showing camels at Kilwa in Jordan; evidence from the temple of Hureidha in the Hadramaut; an inscription of Sargon II acknowledging a gift of camels from Arabia, and so on—does not get us back beyond the eighth century B.C. in Arabia, it is always from Arabia that the camel-riding hordes of Bedouin attack their farming neighbours, whether it be the Midianites attacking ancient Israel, or some other tribe invading Mesopotamia, a century or so later. The Assyrians had constant trouble on their main borders with predatory sheikhdoms whose troops rode dromedaries.

However, we can get further back by other means, and the want of archaeological evidence from central Arabia probably means no more than that archaeology has hardly done more than scratch the surface of that country. There is good evidence for domesticated dromedaries used for warfare in Semitic Akkad—for example, in the time of the first Sargon (2400 B.C.)—and the Semitic invaders of Mesopotamia had come from the south.

It was surely the clear advantage which corps of dromedary riders gave to the Arabs that at last induced rulers of powerful city-states to make use of this unpopular beast. For even when the Arab attacks were beaten off, conventional cavalry or chariotry could not pursue the retreating raiders into the desert; horses and wheels could not go where the dromedary could. The urban states still had no use for the dromedary in their agriculture, although in the arid lands of Africa and Arabia, and probably in Central Asia, dromedaries and two-humped camels were already drawing the plough, often— as one can see them even today in the very arid eastern group of the Canary Islands and in Morocco—in joint harness with a mule or an ass. Nor did they yet have any use for camels as milch or meat animals. But their merchants might by this time have realized the value of the pack-camel for use in linking centres separated by sandy desert; as desiccation progressed the ocean of sand was continually rising and spreading. But, far more important, it was only on camel-back that the troops of the civilized states could meet the raiding Bedouin on equal terms; and at last the Assyrian king Ashurbanipal equipped his officers, and perhaps also some of his other ranks, with dromedaries, a policy which resulted in a victory over the Arabs and a return home with so much loot, consisting mainly of dromedaries, that the animals suddenly became abundant all over Assyria and could be bought for a song, though this last fact might be due as much to the ordinary Assyrian's having very little use for a dromedary as to a glut of animals on the market.

Meanwhile both history and meteorology were working in favour of the dromedary: devastation by constant warfare, as well as climatic changes, was causing large areas of intensively exploited land to revert to steppe as irrigation systems were neglected, and to become fit for very little more than camel pasture. The dromedary was, then, more or less forced on the peoples of Mesopotamia both by nature and by the ravages of unbridled human nature.

It was in such conditions that the use of the dromedary spread into Asia Minor and even Greece, all along North Africa and into Spain. Herodotus (i. 80) tells how Cyrus the Persian defeated Croesus of Lydia at the battle of Sardis (546 B.C.) by mounting his men on dromedaries—or perhaps on Bactrian camels, for his country lay within the zone common to both kinds.

When Cyrus beheld the Lydians arranging themselves in order of battle on this plain, fearful of the strength of their cavalry, he adopted a device which Harpagus, one of the Medes, suggested to him. He collected together all the camels which had come in the train of his army to carry the provisions and the baggage, and taking off their loads he mounted riders upon them accoutred as horsemen. These he commanded to advance in front of his other troops against the Lydian horse; behind them were to follow the foot soldiers, and last of all the cavalry. . . . The reason why Cyrus opposed his camels to the enemy's horse was because the horse has a natural dread of the camel, and cannot abide either the sight or smell of that animal. By this stratagem he hoped to make Croesus's horse useless to him, the horse being what he chiefly depended upon for victory. The two armies then joined battle and immediately the Lydian war-horses, seeing and smelling the camels, turned round and galloped off; and so it came to pass that all Croesus's hopes withered away. . . . [1]

[1] George Rawlinson's translation. Everyman edition.

Xerxes had a camel corps; so did the Romans in North Africa and in Asia. And it was at about the time of Cyrus's victory at Sardis that dromedaries were first introduced into Egypt: they had certainly long been familiar to the Egyptians, but they had been either regarded as unclean and so excluded from the economy, or neglected as not particularly useful in that country.

The golden age of the dromedary all over the Mediterranean basin, east into India and north into Persia came with the rise of Islam and the explosion of the Arabs out of their peninsula all over the Old World south of Lyons and west of China. Thereafter its importance began to decline.

The case of the Bactrian camel remains, for the time being, obscure. Archaeological clues to the animal's early history in domestication are few and ambiguous. The most ancient remains of camels from archaeological sites within this animal's natural range in the wild—from Shah Tepe in northern Persia and from Anau in Turkestan—come from levels indicating a date *c.* 3000 B.C. But there is no means of knowing whether these remains are of wild or of domesticated camels. Vadim Elisseeff [1] considers them to be domesticated, in fact cites them as part of his evidence for the growth of stock-raising in that area at that time. After these, in point of time, come the camel-bone finds on Tripol'ye Culture sites in South Russia (*c.* 2000–1400 B.C.). The Tripol'ye people, famous among archaeologists for their magnificent painted pottery, lived by agriculture and stock-raising, and they might well have tried their hand at domesticating the camel. But again there is no way to be sure that the camel remains found on Tripol'ye sites were those of domesticated camels and all the very early material yet found suffers from this same ambiguity.

From about 1000 B.C. there is abundant evidence, pictorial and written, Assyrian, Persian and Armenian, that by then

[1] In Varagnac (1968).

the domesticated two-humped camel was in use over a very wide range of territory, chiefly as a pack-animal. But wherever the zones of Bactrian camel and dromedary overlapped, the Bactrian always gave way to the dromedary, to which it was in most respects inferior. In only one respect was it clearly superior: *Camelus bactrianus* is hardier, stands cold very much better, than its one-humped relative; and for this reason it has survived in use in Central Asia and northern China.

THE AMERICAN CAMELIDAE

As we have seen, the Camelidae evolved in America and migrated into Asia. In the land of its origin this Order of animals produced a very different beast; but it is not, under the aspect of eternity, so very long since the true camels became extinct in America; not, for example, so long that no man ever saw a true camel living in its natural state in that continent. A human cranium of more or less Australo-Melanesian type was found, for example, associated with animal remains including camel bones at Punin in central Ecuador; [1] these remains have not been dated with assurance but go back to the Pleistocene, and similar cranial remains found in the lowest archaeological levels in Palli-Aike and Fell Caves in Patagonia have been radio-carbon dated 8650 B.C. So man and the true camel did once co-exist in the Americas. Still, the true camel became extinct there long before man reached a cultural level at which he might have undertaken its domestication.

Indeed, the North American Indians, although they had in their territories a number of animals they might have domesticated, were still hunting, not herding, when the Spaniards arrived. There were some farmers among them, but no pastoralists. This has been attributed to cultural retardation; but it seems to me that game was so plentiful in their lands, and maize gives such a marvellous return for so little labour,

[1] Bosch-Gimpera (1968).

23

24

25

23. The domestic pigs are all descended from two wild species, the Wild Boar (*Sus scrofa*)—(23) shows his sow with piglets—and/or *S. indica*, his Indian cousin. For some idea of his transformation following domestication compare . . .

24. . . . the statue of a Landrace sow with piglets, from Aarhus, Denmark; and . . .

25. . . . a fine specimen of a Landrace boar. The bristle coat has been entirely eliminated, or rather refined, and the whole shape, size and rate of growth of the animal modified to suit man's purpose.

26

26. In the nineteen-twenties the Berlin and Munich zoos set out to 'breed back' from certain living races of domestic and semi-feral cattle to recreate their ancestor in Europe, the Aurochs. This fine bull was the result.

27

27. Compare this head of a living aurochs, whose kind were hunted in Palaeolithic Europe and West Asia, and domesticated in West Asia by Neolithic man, with . . .

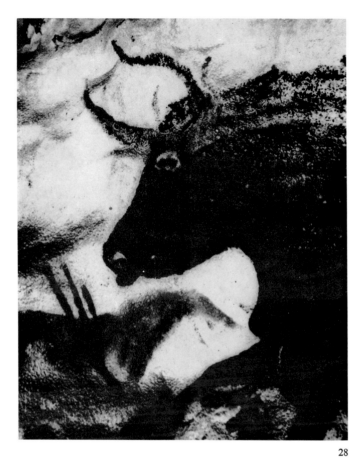

28

28. . . . this painting of a bull's head executed by a great Magdalenian artist on the wall of a cave at Lascaux, France, about twenty thousand years ago.

29

29. Buffalo is a name applied to a number of kinds of wild cattle native to South-East Asia and domesticated in India and, perhaps separately, in China, first as beasts of draught and burden, later as milch animals.

30

30. Yak (Tibetan *gyag*) (*Bos grunniens*), cousin to the aurochs and to the North American and European bison. Native to the high mountains of Central Asia, it was domesticated late in prehistory in Tibet, chiefly as a beast of burden.

31c

31. (a) *Bombyx mori*, the Silk Moth, and cocoons spun by the caterpillar.

(b) *Bombyx* larva spinning its cocoon or pupa case of silk thread.

(c) Silkworm cocoons: a single cocoon yields from 400 to as much as 3,000 yards of silk thread. Selective breeding has produced strains with a higher yield and superior silk to those of the wild *Bombyx*, of which genus several species are domesticated.

31a
31b

that what was missing was the challenge, not the intelligence, to domesticate. Maize, like the potato, is a great fosterer of idleness: a man planting it can easily support his family in return for a total of about fifty days' work in a year. So that the only domestic animal possessed by the Red Indians was the dog; and he had accompanied their remote ancestors when they walked across the Behring land-bridge to populate the American continent, until then empty of man.

The Central American Indians, from Maya down to Aztec, wonderful gardeners and farmers, had no mammalian livestock. They could, I suppose, have undertaken the domestication of certain kinds of deer, but they never did; and there were no other potentially useful species with which to work. They too, of course, had the dog, which in some places was bred and specially fed and fattened for the table.

The South American peoples, or more precisely those of the Andean cultures, domesticated two species: the guinea-pigs, which I deal with briefly elsewhere; and that humpless sub-camel, the Guanaco, *Lama guanicoë*.

There are two domesticated humpless sub-camels in Peru, as there were when the Spaniards first arrived to destroy the great Inca Empire, the Tahuantinsuyu, Empire of the Four Quarters: the Llama and the Alpaca. Both are quite small by Old World standards, an average llama standing about 120 cms. at the shoulder, an average alpaca about 80 cms.[1] No wonder, therefore, that the Spanish chroniclers at first called them Indian Sheep although, oddly enough, the earliest rumours of the existence of these animals refer to them as camels. The llama is kept as a pack animal, the alpaca for its superlative wool.

There are also two wild species in Peru: the guanaco and the vicuña: the guanaco is about the same size as the llama; the vicuña about the same size as the alpaca. Chiefly as a consequence of this coincidence in sizes, the llama was long regarded as a domesticated guanaco, the alpaca as a domesti-

[1] Zeuner (1963).

cated vicuña. But this neat theory has not stood up to modern biological analysis; all the scientific evidence goes to show quite clearly that both the llama and the alpaca derive from the guanaco only, and that the differences between them, e.g. in size and in the quality of the hair or wool, are due solely to modifications in domestication. The llama is the product of selection and breeding for size and strength to produce a better pack-animal; the alpaca a product of careful breeding and selection for fleece. This is important in our context, because such very considerable differences do not emerge and become stabilized in a few generations, and are therefore evidence for antiquity of domestication.

As for the vicuña, it seems that the peoples of the Andean cultures which developed into those great city-states which were finally united in the Inca Empire never did domesticate it. This is curious. True, it is a shy animal and difficult to bring under man's control, yet the Jesuits when they had dominion in parts of South America undertook its domestication and had gone a long way towards success when their banishment from the Spanish dominions by a jealous imperial government aborted the experiment. I should like to advance, with due diffidence, an explanation for the failure of the Andean peoples to undertake the domestication of the vicuña. In the Imperial Inca period the use of cloth made from vicuña hair, obtained by rounding up and shearing wild herds, was confined by strict laws to the royal *ayllu*. (An *ayllu* was a sort of clan, but since in anthropology the words *clan* and *tribe* have very specific meanings, one cannot translate *ayllu* by either.) Royalty, in ancient Peru, was divine. May it not be that the vicuña was deliberately preserved from domestication so that it could provide only the small amount of royal-divine clothes required, and thus maintain a sharp distinction between the garments of royalty and those of ordinary mortals?

Owing to the cutting short of the Jesuits' work with vicuñas, these animals never have been domesticated. Today all the vicuña hair in commerce comes from wild herds which are rounded up from time to time, sheared and released, exactly in the old Inca manner. This, no doubt, accounts for the fact that the price of vicuña cloth in Britain is about £40 a yard, so that the cloth for a man's vicuña overcoat costs about £160. Two things emerge from this account of the vicuña: its handling surely throws light on how, thousands of years ago, sheep and goats were first exploited, just before true domestication; and secondly, that the work which the Jesuits left unfinished should now be resumed. We have for too long been satisfied with the domestic animals which our remote ancestors left us as a heritage; it is time we made some more, and a domesticated vicuña might perhaps be able to exploit not only the very high pastures of the Andes but also those of the Asiatic mountain ranges. It is to the point here that vicuña and alpaca are interfertile and produce fertile offspring.

When the Spaniards first reached Peru—Balboa's first rumours of that rich country were of a land of 'gold and camels'—the domesticated herds of llamas and alpacas were all State owned; the *puric*, citizen-smallholders between the ages of twenty-five and fifty, heads of households and backbone of the country, did not keep these animals and probably were not allowed to. The enormous State herds were guarded by the *llama-michec*, shepherds of the llamas, all of whom were drawn from the labour-conscription class of boys between nine and sixteen years of age. The alpacas were kept for their fleece and, to a very small extent, for meat. The product of the wool-clip was issued by the State to the spinsters—virtually all the married women in the land; the yarn, also State property, was then issued to the weavers, again all women. The *puric* class wore both wool and cotton cloth; vicuña cloth, as I have said, was confined to royalty.[1]

---

[1] This does not mean half a dozen princes and princesses. Since polygamy was the rule, although there was only one *qoya* (queen, usually the Sapa Inca's sister-bride), the royal family were very numerous: they did the work of the higher civil service.

The llamas were used, usually in caravans of several hundred, as pack-animals, carrying goods over the magnificent Inca system of roads and bridges, but also over wild country. The caravan drovers were aided in their work by dogs; each llama carried from eighty to a hundred pounds weight of freight, and a day's move was limited to fifteen miles. Herds of llamas were driven with the Imperial armies on campaign, to provide a ration of meat on the hoof. Llamas were also of importance in the State religion of the Sun, *pillcu llama* ('red' llamas) being used for sacrifices at many rites, including weddings. By the end of the fourteenth century the Inca State's herds of both these animals were so enormous that pasture was running short, not surprisingly when one realizes that the guanaco lives always at very great altitudes, and that llamas and alpacas would not long thrive at much below about 6,000 or 7,000 feet.

As well as these domesticated herds the peoples of the Inca Empire exploited still larger herds of wild guanaco and vicuña, the latter in the manner described above, the former for meat which could be eaten, by ordinary people, only very seldom. Under the laws codified and stiffened by the Sapa Inca Topa Yupanqui (*regnabat* 1471–93) it was a penal offence for any individual to hunt the guanaco, and hunting was confined to an enormous annual *battu* in which tens of thousands of men took part and which was a ceremonial rite as well as an economic act. The wild herds were culled, a proportion killed, and a ration of meat distributed.

So much, then, for the state of affairs in the early sixteenth century. The question now is, when, where and by whom was the guanaco first domesticated? This question cannot be answered with precision; still, there is something to be said about it.

The first agricultural stage in the economy of the Central and South American native peoples is called the Early Farmer period. There is, as far as I can discover, as yet no evidence that the guanaco was domesticated during that period which, in ancient Peru, extends from about 2500 B.C. to about 1000 B.C. But it was not culturally static, quite apart from great regional differences in advancement; throughout that period farming was improving and expanding, and although actual finds of domesticated llama remains have not been made for any part of that period, the finds which have been made suggest that it was, in fact, during the Early Farmer period that the guanaco was first domesticated.

There is every kind of evidence, from plastic representations to vestiges of exquisitely fine woollen fabrics, for all, even the oldest of the pre-Imperial periods, to indicate that the guanaco had been domesticated so anciently that the llama and alpaca had long since assumed their distinctive forms. Perhaps the period on which to concentrate is the Cultist, around 1000 B.C., beginning the phase known as Formative.[1]

The most important culture flourishing in this Cultist period, and the growing-point of South American civilization at the time, was the Chavin Culture, whose type-site is an important temple at Chavin in the northern highlands, and which was first distinguished by the great Peruvian archaeologist Julio C. Tello.[2] Excavations of sites belonging to this culture, in the Viru and Chicama valleys, have revealed a good deal about the economy of the people in question, and from our point of view it is important that they came from a highland home within the habitat of the guanaco. Among the remains found in this area at more than one site were those of llamas. Moreover, from the same period, a small amount of llama-wool fabric has survived. This fabric is not absolute evidence for domestication; the wool of the animals killed in hunting could have been used for spinning and weaving, or wool could have been obtained from wild guanacos, as from vicuñas even today, by rounding up, corralling and shearing

[1] Dates are for Peru only. The same terms applied to Mexico and Central America define earlier dates.
[2] Bushnell (1957).

a wild herd. But that very act is a precursor of full domestication; and it is very possible that the originators of the Chavin
Culture, or at least their ancestors, may have once been
highland nomads living on llama herds, as so many north
Asiatic and North American peoples lived as nomads on
reindeer herds.

It seems to me that we can conclude more or less as follows.
The first stage of domestication of the guanaco, in which the
people lived semi-parasitically on a herd, must have been
accomplished (if we are to account for the marked difference
subsequently developed between llama and alpaca) up in
the *altiplano*, perhaps early in the second millennium B.C.
The second stage, in which distinction between llama and
alpaca was accomplished by selection and breeding in herds
under complete control, probably belongs to the late Early
Farmer and Early Cultist period, earlier rather than later
than 1000 B.C. In short, there may have been a repetition of a
familiar Old World pattern: a descent of a highland people,
with a 'male'-oriented religion and living by animal husbandry, on lowland cultures having a 'female'-oriented cult
and living on plant husbandry. I cannot cite authorities for
these tentative conclusions, still less for the dates suggested;
but they seem to follow from what is known.

To all this it may now become necessary to add that those
tentative dates will have to be very greatly revised. Dr
Richard MacNeish, directing the Ayacucho archaeological and
botanical project in Peru, has made certain discoveries which
suggest a possibly earlier domestication of the guanaco.

A very thick layer of guanaco dung in the Jayhuamachan
Cave suggests that the cave was used as a corral or stable.
Finds made in this cave and radio-carbon dated suggest that
it was occupied by men about 5000 B.C.[1] MacNeish suggests
that in the Andes domestication of the guanaco preceded that
of edible plants.[2] This would confirm the idea of a period of

[1] The stone artefacts found are of the Lauricocha III type.
[2] Buse (1969).

parasitic nomadism based on the guanaco herds. It does not, surely, follow that the Jayhuamachan guanacos were domesticated llamas; numbers of wild animals might have been driven into the cave and penned there until the time came to slaughter them. Still, even that is a beginning and it is quite possible that domestication of the guanaco was accomplished much earlier than I have suggested.

11

Elephants

In this chapter we are concerned with two genera of elephants. *Elephas maximus*, the Indian elephant, has three
geographical races, two of them native to Ceylon, the third
to India, Further India and parts of Indonesia (Borneo and
Sumatra). The original range of this animal was much greater
than it is now. For example, although now extinct in southern
China, it was still to be found there until certainly A.D. 1000
and possibly some centuries later; and there were probably
at least two Chinese races, one of them described as black
with red tusks. Literary records of these animals in China do
not continue later than about A.D. 900, but a passage in
Marco Polo's *Travels* makes me think that elephants may
have lingered on in China, in the wild, a good deal later
than that.

When Marco was on his way from Yunnan to Bengal he
travelled for two weeks through more or less uninhabited
country which was heavily forested and swarming with elephants and other wild beasts.[1] In another passage the King
of Ziamba offers to pay Kublai Khan an annual tribute in
elephants in return for having his country left in peace;
Ziamba seems to have been a part of modern Cambodia. In
the thirteenth century south-west Yunnan was probably
much less deforested than it is now, and it seems probable
that the elephants had not yet retreated entirely into Burma.
It is, at all events, obvious that the range of territory within
which elephants could have been domesticated is vast. Nor
was it confined to the Far East and the Indian centre of the

[1] Marco Polo, XLIII & XLIV. Everyman edition, 1918.

animal's habitat, for a race of Indian elephants formerly inhabited Syria; how they had become cut off and isolated by the extinction of their kind between Syria and India is obscure, but there is no doubt about their presence there. Sir Leonard Woolley found fossil remains at Atchana-Alalakh, which enabled him to date one specimen at 1475 B.C. and another at about a century earlier. There is also a good deal of literary evidence: in the third quarter of the fifteenth century B.C. Thothmes III, the Napoleon of ancient Egyptian history, invaded and conquered Syria. Doubtless by way of relaxation, he went elephant-hunting, a great novelty since there were by then no elephants in Egypt proper. The hunters attacked the elephants at their water-hole and, according to the Pharaoh's biographer,[1] they either found (or actually killed, it is not clear) one hundred and twenty elephants. Thothmes himself engaged in the hunt and was having difficulties with one large elephant when Amen-en-heb cut off its 'hand' (trunk), for which the Pharaoh rewarded him. Those Syrian elephants were still there three centuries later, for Tiglath-pileser I of Assyria (c. 1115–1060 B.C.) also went hunting at that same place, killing ten males and taking four alive.[2] Again three centuries later, this long-isolated colony of Indian elephants was still in being, for Ashurbanipal II hunted them in about 840 B.C., killing thirty. As, throughout the second millennium B.C., and again in the ninth and eighth centuries B.C., they were also being killed by professional hunters working for the ivory trade, it is not surprising that they were declining all the time, and somewhere around 600 B.C. they must have become extinct.

Commercial exploitation of those Syrian elephants was intended to supply the very thriving Syrian and Phoenician trade in articles of carved ivory, a trade which was flourishing during the second millennium B.C. and which was revived by the Syrians in the ninth century B.C. There is no doubt

[1] Amen-en-hab, 1464 B.C.
[2] Zeuner (1963).

that the source of ivory was this colony of Syrian elephants.[1] But for this trade the elephants might well, despite the sporting proclivities of Egyptian emperors and Assyrian kings, have survived long enough to provide a source of elephants for taming and training when the idea of using elephants for work and war finally reached Syria. But no force is more deadly effective in exterminating species than greed for gain, and the demand for ivory was pressing.

So much for the races of Indian elephants. The African elephants present a more complicated picture, zoologically speaking. First of all there is *Loxodonta africana*, which has two geographical races: the gigantic East African, tropical race, which has never been domesticated; and a smaller race, native across the continent from the Sudan to the Atlas where it is now extinct. This is the African elephant which interests us in the context of this book; there are several other African species (unless they too are only geographical races, it depends on which zoologist you follow), but it is the northern race of *Loxodonta*, standing from seven to eight feet high at the shoulder, which is what I will call 'the African elephant of history', the only one domesticated until early in the present century when a small race of Congolese elephants was domesticated for lumbering work in the Congo forests.

In prehistoric times this northern race of African elephants —or perhaps it was yet another race—was native in the Nile Valley. It was probably still there in just pre-Dynastic and early Dynastic times, since there are a number of representations of elephants dating from those times. But the animals must already have been retreating southward, and they became extinct in Egypt proper before the earliest historical epoch; not, however, in the Sudan, where they survived into the nineteenth century when (1868) five of them were shot near the headwaters of the Baraka River.[2] This was the source

[1] Barnett (1939).
[2] Zeuner (1963).

from which the Ptolemies drew their war elephants for taming and training. As will appear, elephants were domesticated
only in that sense; breeding in captivity was not, as in the
case of other domestic animals, relied on.

This brief introduction will, I hope, serve to correct a
number of misconceptions and myths about African and
Indian elephants:

i. 'African elephants cannot be domesticated.' Wrong;
   they can and repeatedly have been, both by the Egyptians
   and by the Carthaginians, as well as in our own time. It is,
   however, true that the largest race of *Loxodonta* has not
   been domesticated.
ii. 'African elephants are bigger than Indian elephants.'
   This is true only of the large tropical race. Indian elephants were bigger than the 'Libyan' elephants domesticated by the Ptolemaic Egyptians, and by the Numidians
   first for Carthage, then for Rome.
iii. 'Elephants do not breed in captivity.' Wrong; they do.
   But see below.

Although both Indian and African elephants are enormously strong, they are not economically efficient as domestic
animals. It will be easier to discuss this in terms of the familiar
Indian elephant, but it is tolerably certain that the arguments
apply to the African ones as well, though the race of domesticated African elephants is extinct.

A domesticated animal, to be profitable, must breed
quickly. Elephants do not reach breeding maturity until
they are sixteen years old; do not conceive more than once
in three or four years; carry their young between eighteen
months and two years; bear only a single calf; suckle it for a
year and keep it at the mother's side for five years, so that her
own work is also hampered; the calves cannot be put to work
until they are fourteen. In this respect, therefore, elephants
are, as domesticated animals, thoroughly unsatisfactory.

A domesticated animal, to be profitable, must give a full

day's work in return for its feed. Elephants consume such vast amounts of food, and still use up so much energy in supporting their huge bulk, that they can work hard for only two or three hours at a stretch; again, most unsatisfactory.

There are no particular difficulties in catching and taming elephants: with the rare exception of the solitary or 'rogue' male, they are tractable. Zeuner (1963) reviews all the methods commonly in use and then points out that all of them were well within the powers and skills of Palaeolithic man, let alone Neolithic man. It is, in fact, probable that those means were invented by Palaeolithic man and used by him against elephants, or at all events mammoths, as well as other animals, although with the object of killing, not taming. Pit traps of the kind still used to capture elephants in some parts of the world were, according to André Varagnac (1968), probably in use by the people who made Abbevillian- and Clactonian-type flint tools and weapons more than half a million years ago. The creatures who once trapped elephants around Torralba in the province of Teruel, Spain, were pre-human hominids, not yet *Homo sapiens*.

It is at least very probable that this whole business of trapping large animals is of tremendous importance in the history of technology, because it was in constructing traps, other than pit traps, that man first used the elastic power of bent wood as a spring, to 'spring' traps, and subsequently to make the bow. Now the use of the energy latent in organic, lignified, vegetable tissue is an enormous step forward from the use of muscle power only; and it leads in due course to the use of the energy that can be stored by bending a piece of tempered metal, i.e. to the metal spring; and from there to the whole notion of machines. All the forms of energy available to man before the nuclear age can, in fact, be traced back to two very early Palaeolithic discoveries: the control of fire, and the use of the energy which hunters learnt to 'store' by bending a piece of springy wood.

But if Palaeolithic, Mesolithic and early Neolithic men

could and did trap elephants, they could not have domesticated them in any sense of the word. They could not have taken even the first step of nomadic parasitism, in the manner of the reindeer men, because you cannot live nomadically on a herd of elephants; the time-scale of their lives, their size, their habits are all quite unsuitable. In other words, man cannot have domesticated elephants until he himself became sedentary, that is until arable farming and plantation farming enabled him to settle permanently, or at least for a very long time, in one place. This, then, means that such domestication cannot have been earlier than late Neolithic in India or in Africa.

But in view of what appears above, why trouble to domesticate such an animal at all? There are two answers: elephants can, despite their disadvantages, carry heavier loads than any other beast of burden, and do work that no other animal can do, for example logging and other kinds of work we now do with mechanical cranes and bulldozers. Secondly it was discovered that they could be (although in practice they often were not) formidable in battle as super-heavy cavalry, each animal carrying several fighting men instead of only one.

Although used in war for many centuries, tamed elephants seem as often as not to have been as great a danger to their own side as to the enemy. When, in Gustave Flaubert's *Salammbô*, the Carthaginian Suffete Hanno attacks the revolting and threatening (because unpaid) mercenary army of the Republic with a force of war elephants, Spendius, the Greek slave who is the brains of the rebel force, puts the great beasts to panic and disastrous flight by driving pigs daubed with bitumen and set on fire towards their advancing ranks. Sainte-Beuve, in his savagely hostile study of the great novelist's book, sneered at Flaubert for this *outré* 'invention'.[1] But it was no invention; that ruse was employed at the battle of Megara.[2] The truth is that elephants as war animals were

[1] *Nouveaux Lundi*, Vol. IV.
[2] Polyaenus, *Strategemata*.

thoroughly unreliable, which is why, after a short trial, the Roman army rejected them.

Europe first became familiar with domesticated elephants in the Roman Circus. The Romans were introduced to them by both the Greeks, or rather Macedonians, and the Carthaginians, in that order. The Carthaginians were using them in war from a little after *c.* 300 B.C., and they made their first massive use of elephantry at the battle of Annone in Sicily in 262 B.C. during the First Punic War, where Hanno fielded sixty trained fighting elephants. The difficulties of transporting the huge animals from Africa must have been considerable, and the elephants proved a failure; for (see page 147) this was not the first time Roman infantry had to stand against an elephant charge. At a battle in Africa (255 B.C.) the Suffete brought one hundred elephants into the field; this time they prevailed, for the Romans under Marcus Regulus were forced to attack the mercenary infantry across the front of the elephant corps, which broke and routed them. This encouraged Hasdrubal, when it became possible to carry the war into Sicily again, to land 140 war elephants on that island. But a year or two later, again in Sicily, the Carthaginian reliance on elephants once more led to their downfall:

In the summer of 251 B.C. the Consul Gaius Caecilius Metellus achieved a brilliant victory under the walls of Panormus. These animals [elephants] which had been imprudently brought forward, were wounded by the light troops of the Romans, stationed in the moat of the town. Some of them fell into the moat, while others fell back on their own troops, who crowded in wild disorder along with the elephants down to the beach, that they might be picked up by the Phoenician ships. One hundred and twenty elephants were captured . . . [1]

Despite what they had seen, the Romans thereafter took to the occasional use of war elephants, although without

[1] Mommsen (1853–6).

enthusiasm; and they soon gave them up as more nuisance than they were worth. Like those of Carthage, the Roman supplies of wild elephants for training came from the Numidians whose forests were full of these animals. But who taught the North Africans to domesticate wild African elephants for war? Curiously enough, the Egyptians.

After the death of Alexander the Great (323 B.C.) one of his generals, Ptolemy, was appointed satrap of Egypt, not assuming the royal title until seventeen years later. In 321 Perdiccas, regent of Macedonia, attempted to invade Egypt with a force that included a corps of Indian elephants; but, having failed to force the Nile at Pelusium, his soldiers mutinied and killed him, and Ptolemy obtained possession of his elephants. Consequently the first fighting elephants to be seen in any part of Africa were Indian animals, domesticated specimens of *Elephas maximus*.

If he was to have the permanent corps of elephantry which he wanted, Ptolemy, or rather his successor Ptolemy II, had to find a source of supply, but it was not practicable to bring enough elephants from India. As a result of this dilemma the African elephant was domesticated for the first time, for Ptolemy made arrangements to have elephants trapped in the Sudan and, using two of the Red Sea ports, carried them to Egypt in specially built transports. Since the Greeks and Egyptians knew nothing about the craft of trapping, taming and training elephants, the king hired experts in India to teach them, and probably a sufficient number of them to mount and control the elephants in battle, at least at first.

The third Ptolemy, Eugertes, fought a number of successful wars against his Asiatic rivals among the Successor States, both sides using elephants, and by this means he captured Indian elephants to add to his Africans. It was from this king also that the Carthaginians received their first war elephant; Zeuner [1] thinks, on numismatic evidence, that

[1] (1963) relying on Gowers & Scullard (1950).

this animal was one of the captured Indian elephants. It was doubtless with the help of Ptolemy Eugertes that Carthage also received her first Indian *mahouts*, elephant-catchers and trainers. This gift inspired the Cathaginians to set their Numidian subjects to the catching and training of African elephants, with the consequences already noted. So, tracing back from the first domestication of African elephants a little after 300 B.C., we are led to Alexander the Great's Indian elephantry.

In the year 327 B.C., having occupied the Eastern Provinces of the Persian Empire following his victory at the battle of Gaugamela in 331 B.C., Alexander the Great invaded the Punjab. Having crossed the Indus and gained the alliance of the rajah who ruled the country between that river and the Hydaspes, and having seen their first few war elephants at Gaugamela, in the spring of 326 the Greeks advanced to the Hydaspes where the aged rajah they called King Porus was waiting to bar their advance with 30,000 men and 200 war elephants. Arrian describes this battle and the stratagem which Alexander used to win it; all I need say here is that, as so often happened, the elephants were more dangerous to their own side than to the enemy, retreating, as Arrian puts it, 'like ships backing water'.

It was, then, at Gaugamela and at the battle of the Hydaspes that Western man first came into contact with domesticated elephants. Alexander immediately adopted their use, which accounts for their subsequent employment by the satraps in their internecine wars. He no doubt considered his victory as due to the superiority of himself and his men as soldiers, rather than to any failure on the part of his opponent's elephantry. As we have seen, Perdiccas used elephants against Ptolemy two years after Alexander's death. Later still, Seleucus I of Syria gave the rajah Chandragupta a whole province in exchange for 500 war elephants, and with them won the battle of Ipsus in 302 B.C. The Syrians even tried chariots drawn by a team of four elephants, but so clumsy a device can only have been used ceremonially. As an indirect result of all this the Romans got their first taste of an elephant charge, for King Pyrrhus of Macedonia used elephants against them—and defeated them—at Herakleia in 280 B.C.

It was, then, because the domesticated elephant had long been in use in India that the African elephant was domesticated. I continue to use the word domesticated, although in fact the animals were all caught in the wild and tamed; breeding in captivity was relied on only where and when some special attribute, such as albinism, was required for religious purposes. With a short-lived or intractable animal this method of catching and taming would not have been sound practice. But as elephants are quickly trained, have a useful working life of about forty years, and live to be seventy or more, it is perfectly practicable.

So a last question remains: where and when did the Indians first begin this practice of taming captured elephants instead of killing them?

There is abundant archaeological evidence that elephants were in use for carrying burdens and for riding in several parts of India from about 500 B.C. onwards. The earlier pieces—coins and terracotta toy elephants—are from the Punjab, and it is evident that this was the elephant-centre of India. But there is much older evidence, and if there is a gap of over a thousand years between this early evidence and that of 500 B.C., it can be due only to the fact that the archaeologists have still much to dig up in India.

Fifteen cylinder seals bearing figures of elephants, found in excavations at Mohenjo-daro, metropolis of the Indus Valley civilization, are dated between 2500 and 1500 B.C. Of these, six are of particular interest in our context because on them the elephant is shown wearing a covering cloth. One should in fairness point out that this has been denied by some authors who argue that the 'fold' from which this covering is deduced is a skin fold; but Zeuner (1963) has shown very clearly that the fold on these six figures, extending down-ward from the middle of the back behind the neck, to just behind the foreleg, can be nothing but such a cloth. Now, an elephant wearing a covering cloth is, without question, a tamed elephant.

The date of these six seals cannot be fixed very precisely; it appears that they can hardly be earlier than 2500 or later than 1500 B.C. Then, too, we do not know how long after the elephant had been domesticated the seals were made. One can say only that elephants were first tamed and put to work in the Indus Valley probably in the third millennium B.C. certainly not later than the first half of the second.

12

Fur and Flesh

A number of animals much smaller than those with which
we have been dealing have been domesticated from time to
time; most of them are rodents, but now we are engaged in
domesticating more small fur-bearing carnivores. In the past
the usual reason for such domestication has been to provide
meat; in our own time, although meat is as much needed
as ever it was, that very powerful engine in our society, the
'profit motive', has turned attention to fur-bearing animals.
The demand for fine furs, and the money to be made in
supplying them, has indeed been the only force strong enough
in modern times to promote the domestication of animals
left wild by our ancestors; the only force that could persuade
us to add anything, in the way of domesticated animals, to
the heritage left us by Neolithic man, even though half the
world's population is starved of animal protein food and
there are still several animals which, domesticated, might
help to supply them. Modern man has not done well in this
field; his achievement has, in fact, been negative. The elk
was formerly domesticated for meat and milk, and elk stags,
being faster than horses, were ridden by the couriers of Karl
XI of Sweden in the late seventeenth century; but the animal
has been lost to domestication.[1] Many species of antelope
were domesticated in the past, and so were fallow-deer.
Gazelles were domesticated for meat in Egypt at a very

---

[1] According to Zeuner (1963) the Yakuts were riding elk, also
used as a draught animal, into the nineteenth century. As a Yakut
on an elk could always outride the Czarist police on horses, the
Russian government prohibited the keeping of domestic elks.

early date, and probably herded like goats. Even Ibex seem
to have been domesticated. All those animals have been lost
to the stock-breeder.

RABBITS

Everyone in Britain who is over forty can recall a time when
it was impossible to go out for an evening walk in the country
without meeting with the pretty spectacle of rabbits feeding
and at play near their warrens; a time when any countryman
could, at the cost of an evening stroll with a gun and dog,
bring home a couple of brace of rabbits for the larder; when
one could not drive along a main road away from towns without seeing the pitiful sight of hundreds of rabbits killed and
spatchcocked by motorcars; and when, finally, Australia
suffered such a plague of these swiftly proliferating animals
as did serious damage to her economy. The pandemic of
myxomatosis which nearly exterminated the rabbit species
changed all that. But before that disease, spread deliberately
by farmers tired of seeing their crops eaten by rabbits, had
reduced the rabbit population to a few thousands, it was
difficult to imagine a Europe without rabbits, and one was
apt to think of them as native from time immemorial all over
Eurasia.

But this was not so: a little more than two thousand years
ago the rabbit was a novelty in most of Europe; it reached
the north as a domesticated animal some centuries before it
arrived as a wild one, and largely as a result of its domestication. Aristotle had never heard of it, nor had Greek writers
who followed him. In fact the first of them to mention the
rabbit was Polybius, in the mid second century B.C., and it is
significant that he called it *kyniklos*, an obvious loan word,
the Latin *cuniculus*. In other words, it was through the Italians
that the Greeks first heard of rabbits, which means that the
rabbit came from the far west; in fact, the Latin word is
probably rooted in ancient Iberian; it meant also underground

burrows or passages, but it did not come to mean that until after the rabbit had demonstrated its burrowing powers in Italy.

The Romans, like other civilized people before them, first came across the rabbit in Spain. Heyn (1888) was of the opinion that rabbits came to Spain with the Iberians themselves, from Africa. They may have done, of course; but Zeuner, a better authority, says that they were probably restricted to the Iberian peninsula after the last Ice Age. Both could be right, because Tangier, after all, is so near to Gibraltar that if southern Spain was spared the worst effects of that glaciation, so also must north-west Africa have been. At all events, what this means is that although the rabbit's range before the last Ice Age may have been much wider—fossil remains suggest that this was so—the Iberian peninsula was the only part of Europe where the species survived the ice. Then, after the retreat of the glaciers, the rabbit population built up again. Confined to the peninsula and perhaps south-west France by sea and mountains, and despite an invasiveness forced on it by the fact that a single female may produce seventy young per annum, the rabbit required the help of man in order to extend its habitat. And that help it began to receive after the Romans met with it in Spain; and as a result, in a few centuries all Eurasia was overrun with rabbits.

In different environments they became differentiated into a number of geographical races, varying in size, colour and other attributes, but they all belong to the single species *Oryctolagus cuniculus*.

It is curious that Europe should have had to wait so long for rabbits, when one considers that they were known to the Phoenicians in Spain from about 1100 B.C. and that the name which Spain still bears, given to it by the Phoenicians, means something like *Rabbit-land*, although that is a very free translation. The Phoenician seamen who landed in Spain towards the end of the second millennium B.C. noticed that the land was mined and burrowed by very numerous small, furry animals which, at least in their burrowing habits, were something like the *shephan* of their own country. This led them to name the newly discovered territory, the most westerly land of the known world, *ishephanim* (hyrax-land), which name becomes, by corruption, Hispania (modern España). Perhaps those men of Tyre and Sidon made no use of the newly discovered animal, so that, if it was not already in North Africa, it was left to the Phoenician colonists in Africa, the people of Utica and the Carthaginians, to introduce the rabbit into what are now Algeria and Morocco.

If we suppose that the rabbit had never been domesticated in Spain, then the man responsible for its domestication was Varro, and the thing was accomplished in Italy. The idea of keeping and breeding these new animals in captivity was suggested to him by the existence of *leporaria*, which by then were quite common in Italy: the word could be translated as 'hare-aries', extensive walled plantations planted with grass and shrubs, in which the Romans kept hares. These were not domesticated, for they lived naturally; but, being under control, they could easily be caught and killed when wanted for the market or the table.[1] The hares were so far from being domesticated that they did not even breed in those enclosures. As Zeuner (1963) points out, *cunicularia* would, unlike *leporaria*, need to have walls founded far below ground level because, whereas hares do not burrow, rabbits would soon have dug themselves out. They did, anyway, after *cunicularia* had been established at Varro's suggestion; and the escapers were the ancestors of all the wild rabbits in Italy.

The keeping of rabbits in such enclosures, although they, unlike the hares, did breed, cannot be considered to have been domestication, which involves a very much more advanced measure of control. The argument, used by Zeuner, is taken from Nachtsheim (1949) and it comes to this, that in such rabbit-enclosures as Varro suggested and the Romans

[1] Varro, 116–27 B.C. *De re rustica.*

used, '. . . selection would work in favour of wild characters and against domestication since, owing to the necessity of catching or hunting the rabbits, the quickest and "wildest" would have the best chance of survival whilst the "tamer" ones would be killed off.'

Rabbits were kept in this manner, that is in enclosed warrens, not only into the Middle Ages, but much later. Such rabbit gardens were common in France in the fourteenth century, in England and Germany into the seventeenth century. For obvious reasons they were often made on islands, usually in lakes or rivers, or just off-shore. Throughout this period of seventeen centuries, then, Nachtsheim's rule was at work; no attempt was made in the enclosed warrens to select mutants or variants, and the stock remained wild. The rabbit would never be domesticated merely by being kept in captivity. But meanwhile the little beasts were escaping to establish wild rabbit populations in France, Germany and England. It is possible that enclosed warrens were created in the grounds of Romano-British and Romano-Gallic villas before the fourth century, and if so then probably the first wild rabbits to be seen in northern France and Britain appeared there in Roman times. But this is by no means certain and there were unquestionably parts of Europe where wild rabbits did not put in an appearance until long after true domestication had been accomplished, and domesticated rabbits were being kept in monasteries all over the Continent. By the twelfth century this was the case, yet the wild rabbit is not reported from, e.g., south Germany until the fifteenth.

Which still does not tell us where and when real domestication began. Nor why; and the why of it is rather singular. It begins, like the domestication of dormice (see below), with the Roman taste for luxury foods.

The Romans, after they had established rabbit gardens in the manner described and acquired a taste for dishes of rabbit in their cuisine, were soon carrying their gastronomic vulgarity to extremes, as they always did. The Trimalchios of their, as of our, world, thwarted in their passion for accomplishing conspicuous consumption in terms of mere quantity by the limitations of even the most gluttonous appetite, turned to fantastic dishes as a means of spending more money on food. In the case of the rabbit this began with the eating of sucklings and ended with the eating of immature rabbit foetuses—known as *laurices*.[1] This distasteful practice was carried on by the early monastic establishments as soon as strict rules began to be relaxed, and for a very odd reason: adult rabbit might be meat and therefore forbidden on days of abstinence; but rabbit foetuses were considered to be a sort of fish, and could therefore be served on Fridays and even in Lent. Embryologists tell us that the foetus in the womb of a mammal does pass through a sort of fishy stage so maybe this piece of sophistry has some scientific backing. And when one considers that beaver was eaten on days of abstinence on the grounds that since it lived in water it must be a fish, the practice of eating rabbit-foetuses is not so surprising.[2] Not that all churchmen approved: Gregory of Tours, for one, deplored the eating of rabbit foetuses in Lent.

Such, then, was the very improbable cause which gave rise, probably in the sixth or seventh century, and possibly earlier, to the real domestication of the rabbit: complete enclosure, controlled breeding; selection of mutants, and line-breeding. In the following centuries black rabbits, white rabbits, piebald rabbits, and rabbits with specially fine fur were isolated into domesticated races and strains. After the sixteenth century rabbits which were the outcome of centuries of selective breeding were valued not only as a source of meat, but also as a source of fur.[3]

[1] Heyn (1888).
[2] Zeuner (1963).
[3] To forestall an obvious criticism of this account, it should be added that the coney or rabbit of the sixteenth- and seventeenth-century English translations of the Bible is a mistranslation; the word translated coney really means hyrax.

THE SILVER FOXES

It seems that fox-farming did not begin with modern fur-farming, and that there was probably a prehistoric domestication of the fox. The fox was hunted in Palaeolithic Europe, not only for its flesh but also for its fur; in some Swiss Neolithic lake-dwellings fox remains are more common than those of dog, and certain deformations of fox-bones found on such sites suggest that the fox had been domesticated.[1] Nor are bone deformities the only evidence, for these foxes, although of the same species, are smaller than the wild foxes of the region at the time in question, and in the first stages of domestication in primitive conditions animals tend to be smaller than their wild congeners. But if the Swiss Lake Dwellers did, in fact, take a step towards the domestication of the fox, it had no future; for, thereafter, there is no sign of domesticated foxes for many centuries.

Late in the nineteenth century Canadian fur-trappers, or perhaps just one of them who communicated the idea to a few friends, conceived the idea of keeping summer-trapped foxes—the white, silver, and 'blue' arctic foxes—alive, penned and well fed, until mid-winter, and only then to kill them and send the pelts to market. The prices fetched by furs on the London market, at that time the most important in the world, varies with the quality of the pelts. The summer fur of any fur-bearing animal is not nearly so rich and full as the winter fur, a natural provision for dealing with the seasonal changes in temperatures. So some of the trappers, by keeping the foxes alive until winter, were able to make a great deal more money than they had done. Furs of finer quality began to reach the London market, and very high prices were paid for them, the record for the year 1900 being £580 for a single skin. It soon began to be known that certain trappers were making fortunes by this new means.[2]

[1] Hauk (1950).
[2] Aitken (1963).

That was the beginning of domestication: two men, Sir Charles Dalton, and R. T. Oulton, set about investigating the practicability of farming *Vulpes fulva* and *Alopex legopus*, that is to say of breeding them in captivity. Their initial trials were entirely successful, an industry was founded, and by 1910 it was expanding and showed such promise of large profits that in 1913 a first-class pair of breeding foxes sold for $35,000. A side-effect of this boom was curious—a lot of option-gambling in unborn cubs. However, the First World War and its aftermath steadied the new industry, which by 1927, still expanding, had settled down.

Thus two species of arctic fox were domesticated between 1900 and 1927. Change in fashion, and the later rise of mink farming, has led to a great decline in fox-farming all over the western world. Only in the U.S.S.R., where the harsh winter climate makes fur more nearly a necessity than a luxury, and where fashion changes much more slowly, is fox-farming still, at the time of writing, an important industry. But another change in fashion can at any time create a new rise in demand for silver, blue, and white fox; and this time the fully domesticated animal is available.

Rabbits, then, were probably the first fur-bearing animals to be domesticated for their pelts as well as their flesh, unless we count the little foxes of the Swiss Lake Dwellers and also ignore the fact that dog-skin and the hides of all domestic animals have been important by-products of domestication since prehistoric times. But then, as a result of success with foxes, a number of other fur-bearing animals were taken into domestication—musk-rat, skunk, pine-marten are examples and, most important of all, mink. Some curious results, analogous to what happened when the rabbit was first introduced, have followed. Farm escapes, and a few cases of irresponsible abandonment of failed farms, have led to additions of new species to the fauna of Britain and other countries: feral minks and feral musk-rats, which are establishing themselves and are perhaps finding ecological niches. I shall have a little more to say about this. Apparently even moles have, from time to time, been 'farmed' for mole-skin, treated in much the same manner as rabbits in Roman rabbit-gardens.

MINK

Canada is the native land of the mink, although the habitat extends into the United States also, and mink is a member of the Alaskan fauna. The best wild minks for fur quality are trapped in the regions of Eskimo Bay and Hudson's Bay. The Alaskan mink is larger and more fertile, bearing up to six kitts at a birth. By about 1920 the prices of pelts of this very beautiful and hard-wearing fur on the London auction market were so high that a few men, with the experience of fox-farming in mind, realized that mink-farming, if it were feasible, would be even more profitable. Trials made in Canada and the United States were followed up quickly in Norway, Sweden and (1928) Britain. There were some fears that mink-fur would never reach prime condition in Britain, because of the mild winter climate. Those fears have turned out to be excessive: apparently British humidity acts on the mink skin much as cold does in Canada.

So completely at home has the mink made itself in Scandinavia and Britain that feral animals have established their species (*Mustela vison*) quite firmly, choosing riparian habitats as a rule, living on fish and, unfortunately for their future in our wild, on domestic poultry of which they are great thieves. As this mink is descended from domesticated hybrids of a large Alaskan race and a race from the state of Quebec remarkable for its fine pelt, it is a tough and fairly fertile animal and has, despite trapping and official controls, a very good chance of becoming a permanent member of our fauna. Minks are now established, at all events, in thirty-one of the English counties. The same thing has happened in Sweden, where about 20,000 wild minks are now trapped every year; in Norway, where the annual catch of wild minks is about 15,000; and in other countries where mink has been introduced. Within a few generations the mutant colours and other attributes of domestication will disappear, and our British wild mink will be dark brown. The large size obtained by selective breeding in domestication is already disappearing in the new wild: a male domesticated mink weighs between 7 and 8 lbs; the largest of the wild ones, out of a sample of 1,000 caught in Britain and weighed by Ministry of Agriculture officials, weighed 3 lbs 10 ozs.[1]

The dangers of such introductions are, of course, very well known. The rabbit in Australia is the usual modern case in point, although there is a classical precedent: according to Strabo, a single pair of rabbits taken to one of the Balearic Islands soon produced such a plague of the creatures that the unfortunate inhabitants petitioned the Emperor Augustus to send military aid, or to evacuate the island and settle the people elsewhere. Feral musk-rats in Britain have been causing serious damage to river banks by mining them. The minks, I suppose, if they multiplied sufficiently, might play havoc with game-fishing; but it is difficult to see what other damage they could do, and there may be an ecological niche for them left vacant by the decline of the otter population.

Mink makes a good example of how domestication works to change certain attributes of an animal, because with so much profit at stake selective breeding has been careful and intensive. By the segregation of mutants of exceptional fur colour and by line-breeding; by rich but correct feeding; and by a measure of restriction of activity, minks of rare colours—pale fawn, pure white, blue and even pink—have been produced, and the size of the animals increased very greatly. As a result, and for the time being, domesticated minks are economically by far the most important of farmed fur-bearing animals, at least in the West.

[1] Thompson (1968).

Work similar to that done on minks in the West has been done on sables in the U.S.S.R.

GUINEA-PIGS

Cavea porcellus is a rodent. Unknown in the Old World before about 1550, thereafter it began to turn up in all the countries of the Spanish and Portuguese empires. It was imported more as a curiosity and a pet for children, like the Golden Hamster in our own time, than for any useful purpose. It reached Spain before the end of the sixteenth century, but it was from West Africa that it was introduced into England—probably in the seventeenth century—whence the vernacular name, Guinea-pig.

When, in the nineteenth century, biologists and pathologists began to need laboratory animals, the guinea-pig, cheap because prolific, tractable, easy to feed, house and handle, had the melancholy honour of becoming the laboratory-animal *par excellence*, and persons of sensibility will shrink from contemplating what this harmless little creature has suffered on man's behalf. This role has become, as they say, proverbial, for we now speak of even a human being on whom anything new is to be tried as a guinea-pig. So, with the rise of experimental pathology, began a second episode in the domestication of the cavy.

The first started in pre-Inca Peru, either early in the history of the several urban States subsequently united by force of Inca arms into the Tahuantinsuyu, or of one of them; or again, much more likely, in one of the pre-urban cultures based on agriculture which ultimately gave rise to those nations. That the domestication of the cavy was long pre-Inca is established not from archaeological evidence—there is very little; I will come presently to what there is—but by looking at guinea-pigs. For by the time that Europeans came upon the scene the domestic cavy was very unlike the wild one, in size, in fur colours which included piebald, and in

fur texture. The reader, bearing in mind what has been said
above about minks whose domestication is not yet a century
old, will be inclined to raise an eyebrow at this point. And
it may be that, in the light of our experience with minks, we
shall have to revise some conclusions about the significance
of certain morphological changes for the time-scale of
anciently domesticated animals. On the other hand, we have
the enormous advantage of knowing, what the ancients did
not, some genetics.

Domestication of the cavy must have been accomplished
in much the same manner as that of the rabbit many centuries
later. Its object was meat. With only the slow-breeding llama
as a farmyard animal (except some tamed wild-fowl) and
some hunted deer, the Andean peoples were short of meat-
sources. By Inca times the cavy was the principal source of
meat for the Inca Empire; even today it remains one of them.

What may be new light has been thrown on the beginnings
of the domestication of this little animal, by Dr Richard
MacNeish's excavations at Jayhuamachay in 1969. At the
back of this cave, occupied by Palaeolithic men whose culture
has been dated to *c.* 5000 B.C., and which had also been used
as a corral for guanacos (see Chapter 10), he found abundant
fossil remains of cavies. This does not, of course, mean that
they were already domesticated, as was claimed by Peruvian
newspaper accounts of the finds; it probably means no more
than that cavies formed a part of the diet of Andean man
about 7,000 years ago, which is hardly surprising. On the
other hand, it may mean that a 'herd' of cavies was kept and
fed in a cave until wanted for meat; and that, although far
short of domestication, is a good first step towards it.[1]

CHINCHILLA

During the second half of the sixteenth century Sir John
Hawkins, the famous or, if you prefer, infamous privateer,

[1] Buse (1969).

was founding a fortune by hijacking the cargoes of Portuguese slavers, and selling them in Spanish colonies. The trade broke the laws of the Spanish Imperial government, but the Spanish colonists were happy to be able to buy slaves at less than the ordinary market prices. In the course of his trade Hawkins saw many lands and he had a remarkable gift for exact observation, for recording what he saw. Here is a short passage from his writings:

Among them they have little beastes, like a squirrel but that he is grey; his skin is the most delicate, soft and curious furre that I have ever seene, and of much estimation, as of reason, in Peru. Few of them come into Spain because difficult to come by, for that princes and nobles lay in wait for them; they calle this beaste chinchilla, and of them they have great abundance.

There are more than one species, but two native to the Andes seem to have been involved in the domestication: *Chinchilla lagetis* and *C. vischaca*. Like the guinea-pigs, they were domesticated by pre-Inca Andean people, but not for their flesh. The fur was either sheared or plucked and made into a rare cloth—it is, as Hawkins noticed, very long and silky. I do not know how this was done; surely the staple would be too short for spinning? But then, the Andean peoples were as remarkable for their extraordinary manual dexterity and skill as for their technological poverty, and on their primitive looms they wove cloth finer, with more threads to the inch, than any cloth made in the world until modern times. Nor can I discover whether the Incas were using chinchilla pelts for furs, or whether the Spaniards were the first to do so.

At all events, in due course export of pelts to Europe began, and they were not at first dear; on the contrary. As late as 1837 chinchilla pelts could be bought on the London market, by furriers, at $2.16 or 18*s*. the dozen.[1] Such a price implies that wild chinchillas were being slaughtered at an alarming

[1] Clarke (1961).

rate; and that this was indeed the case is apparent from the trade in fur, for the decline of wild chinchillas in the Chilean Andes rapidly forced up the price of pelts, a process assisted by a fashion for chinchilla coats. The following little table [1] will show what was happening:

Year	Number of skins exported from Chile	Price per skin
1899	433,463	$3 or 12s.
1905	247,836	$9 or £1.16s
1910	153,863	$60 or £12

This process of over-exploitation continuing, the Chilean Government put an embargo on chinchilla trapping in 1918. Some illicit trapping continued, for in 1927 there were still chinchilla skins to be had on the London market at about £30 a pelt; still, the chinchillas were given a chance to recover.

Even before 1918 there had been some attempts, all of which failed, to domesticate chinchilla. The special difficulty was to acclimatize them to altitudes lower than the very great altitudes of their natural habitat.

In 1918 an American mining engineer working in Chile, M. F. Chapman, began to take an interest in the problem of domesticating chinchilla. With a licence from the Chilean Government he trapped animals at high altitudes, kept them in suitable enclosures or rather cages where they could live and breed, and between 1918 and 1923 brought them down in stages, with long halts between moves, to sea level, meanwhile studying their diet and habits. So successful was he that when he had to return to California, taking his chinchillas with him, they were able to live and breed at the relatively low altitude and in the warm, dry conditions of his home.

Others followed Chapman's lead; the keeping and breeding of chinchillas began, and by 1930 a good breeding pair cost

[1] *ibid.*

$3,200, although the regular sale of chinchilla pelts from farms did not begin until the 1950s, and this extraordinary rise in prices of domesticated chinchillas was due to nothing more than the prospect, not to any certainty, of profit. At a preliminary New York chinchilla-pelt auction in 1946 pelts fetched up to $56 each; but it became clear that a great many of the farmed pelts were inferior to the wild ones, and that only the best were good enough. Since then, the general standard of quality has been improving and there can be little doubt that the domesticated chinchilla is here to stay. Mutant colours have already appeared; breeders are engaged in suppressing the grey and bringing out the blue, and no doubt the story from now will be very like that of the minks.

DORMICE

If we exclude insects, the dormouse is the smallest animal ever to be domesticated by man. Its domestication was a product of Roman gluttony: the Romans of late Republican and Imperial times had a somewhat *outré* taste for exotic foods; or, rather, the class of the very rich had. In this case the victim of their gastronomic lust was the 'Fat' Dormouse, *Glis glis*, and from the Roman glutton's point of view, the fatter the better. What they were exploiting was the animal's build-up of fat on which to live during its winter hibernation.

Glis glis is native to all continental Europe and western Asia. It lives in trees on a diet of acorns and all kinds of nuts. At first it was hunted, but by about 100 B.C. the suppliers of the Roman luxury market, or the stewards on great estates, had established *gliaria*, enclosures planted with suitable trees and in which the dormice were given a rich diet of all kinds of nuts, chiefly acorns, walnuts and chestnuts, so that they grew much fatter than they had a chance to do under natural conditions. There is a wonderful 'Edwardian' touch about Ammianus Marcellinus's account of Roman millionaires placing scales on their tables to weigh the dormice served to their guests, and having notaries in attendance to verify and record the weights attained.

We are apt to think that the revolting practices of 'factory farming', such as keeping calves tied or penned motionless, and in the dark, to obtain white veal from larger and heavier and so more profitable animals than would yield white veal in ordinary farmyard conditions, are all modern. Nothing could be further from the truth; they are no more than industrializations of old practices. Stall-feeding of cattle is a very ancient south European practice; the ancient Egyptians kept hyaenas, bound so that they could not move, on their backs, and forcibly stuffed them with special foods to make them very fat. The same people invented the goose-stuffing torture which produces *pâté de foie gras*, fatty-degeneration of goose-liver. The Roman dormice-farmers kept their most promising animals in small earthenware vessels called *dolia* and, to fatten them excessively, over-fed them in such conditions that they could scarcely move.

13

Domesticated Fish

My theme in this and an earlier book has been man's work in easing his lot on earth and enriching his societies (and so his life), by bringing food- and raw-material plants, and animals for food, raw materials and muscular energy, under his control and transforming them so that they suit his needs. In a sense, domesticated plants and animals are artefacts. In an earlier volume [1] I tried to give an idea of how, when and where man, no longer willing and perhaps no longer able to take a chance of finding food and material in the wild, ceased to be a collector or gatherer of food and became a producer. In this volume we have been concerned chiefly with the story of how, from being a mere hunter of animals for meat and hides, bone and horn, he became a stock-breeder and herdsman, controlling, using and slowly transforming a number of animal species. Today, in all but a few jungly backwaters of the world, man hunts on land only by way of recreation; and it is a curious fact that the very men who have least need of hunted food are the ones who most readily turn to hunting in their leisure, that is, the rich. There is a pleasure in the hunting and killing of animals which must be a heritage of our long Palaeolithic past.

All that, however, is true only of man on land; on the seas and lakes and rivers we are still hunters. Apart from the fact that the gear we use for fishing is more complex, we are no forwarder on the waters than were our fishing ancestors of 20,000 years ago. The fish-hook and the fish-net are about 25,000 years old. The fish-hook is a Magdalenian invention. [2]

[1] Hyams (1971).
[2] Breuil & Lantier (1959).

We are still using hooks and nets and fish-traps, and we have only just begun to take steps towards the control of maritime, riverine and lacustrine fauna, steps which, in respect of land animals, we began to take 5,000 or 10,000 years ago and completed, but for a last few paces, over 5,000 years ago.

Nevertheless a number of first steps towards the domestication of useful fishes were taken in the past. I propose to review them briefly and then to go on to what is being done in our own time. For it is in the sea that the next great advance in farming must come, if only because uncontrolled or imperfectly controlled 'hunting' of fish with more and more effective means is seriously reducing fish populations in many parts of the world and even endangering the survival of some species, such as salmon. We are probably approaching the point at which it will no longer be true to say, as we do in the proverb, that there are more fish in the sea than ever came out of it.

Zeuner (1963) allows only four species of fish ever to have been truly domesticated: *Muraena muraena*, variously known as Roman Eel, moray and murena, a species centred in the Mediterranean but found also in other seas; *Cyprinus carpio*, the carp, native to rivers and lakes from the west of Eurasia to the Far East; *Carassius vulgaris*, the Crucian carp which has much the same natural range; and *Macropus viridiauratus*, the Paradise Fish of south-eastern Asia. The same author adds that some modern aquarium fish could be appended to this list; but as purely ornamental fish they are, like the Paradise Fish, somewhat outside the scope of this book. Of the four species in the list the first two were domesticated for food and never had any other role; but the third, domesticated in the first place for food, revealed in domestication such ornamental attributes that it was bred for them, and so, as the Goldfish, became a purely decorative fish.

The most ancient civilizations in Mesopotamia, Egypt and China—and possibly Peru, for the Incas had great fishponds in their gardens when the Spaniards arrived in the sixteenth century—all made attempts to keep and farm fishes. But the remains of their attempts are too vestigial to tell us much about them: we do not know what fish were kept, or whether they bred in tanks or were simply stored alive in those tanks. There were attempts not quite so remote in time as these initial ones. At Agrigentum the Sicilian Greeks used Carthaginian prisoners of war [1] to dig a colossal fish-rearing tank, about fifty acres in extent, which was in use for some time but was finally, it seems, abandoned as impracticable and filled in.[2] From this and probably other Sicilian examples the Romans, at war with Carthage in Sicily, brought the idea of fish-tanks home to the Italian mainland, and there during the first century B.C. many *piscinae* were made, some of them filled with sea-water. Some of these great tanks on the sea-coast were probably rather like those we ourselves are making (*see* page 172). Here is a passage from Philemon Holland's delightful translation of Pliny's *Natural History*, which seems to me to suggest that such tanks, fitted with fish-proof filters or dams of some kind, may have been connected with the main body of the sea. Pliny is writing (ix. 30) about Trebius Niger's observation on octopi, which he calls, or the translator calls, Polypi:

The rest which mine author hath related as touching this fish, may seem rather monstrous lies and incredible, than otherwise: for he affirmed, that at Carteia there was one of these Polypi, which used commonly to go forth of the sea, and enter into some of their open cesterns and vauts among their ponds and stewes, wherein they kept great sea fishes, and otherwhiles would rob them of their salt-fish, and so go his waies againe: which he practised for so long, that in the end he gat himselfe the anger and displeasures of the maisters and keepers of the said ponds and cesterns, with

[1] They would not have been strictly Carthaginian, since Carthage had no national army except for an officer corps, and used only European mercenaries.
[2] *Diodorus Siculus. c.* 20 B.C.

his continuall & immeasurable filching: whereupon they
staked up the place and empalled it round about, to stop all
passage thither. But this thief gave not over his accustomed
haunt for all that, but made meanes by a certaine tree to
clamber over and get to the fore-said salt fish; and never
could he be taken in the manner, nor discovered, but that
the dogges by their quick sent found him out and baied at
him: for as he returned one night toward the sea, they assailed
him and set upon him on all sides, and therwith raised the
foresaid keepers, who were affrighted at this so sudden an
alarm, but more at the strange sight which they saw.

For first and foremost this Polype fish was of an unmeasur-
able and incredible bignesse; and beside, he was besmeared
and berailed all over with the brine and pickle of the foresaid
salt-fish, which made him both hideous to see to, and to
stinke withall most strongly. Who would ever have looked
for a Polipe there, or taken knowledge of him by such marks
as these? Surely they thought no other, but that they had to
deale and encounter with some monster: for with his terrible
blowing and breathing that he kept, he drave away the dogs,
and otherwhiles with end of his long stringed winding feet,
he would lash and whem them; somtimes with his stronger
clawes like arms he rapped and knoked them well and surely;
as it were with clubs. In summe, he made such good shift
for himself, that hardly and with much adoe they could kill
him, albeit he received many a wound by trout-spears which
they launced at him. Wel, in the end his head was brought
and shewed to *Lucullus* for a wonder, & as big as it was as a
good round hogshead or barrel that would take and containe
15 Amphores.

The implication seems to be that sea-fish were at least
'penned' if not actually reared at Carteia. Muraenas were
kept and fattened in salt-water tanks. Some rich men and
women had pet muraenas, and even adorned them with
valuable jewellery such as fin-rings of precious stones. The
millionaire usurer and speculator Marcus Crassus who, as a
result of his immense wealth became one of the Triumvirs,
had just such a pet fish which would come to him when
called to be fed, answering to its name. There is a hideous
and almost certainly unfounded story that muraenas were

sometimes fed on the flesh of slaves. It probably arose from a bit of gossip in Seneca's *De re*, II. 40; see Zeuner (1963).

Muraenas were not simply kept, but bred in these 'stews', chiefly for the luxury market. Roman millionaires indulging in conspicuous consumption entertained enormous numbers at their banquets, and required very large quantities of fish. Muraenas were not the only fish so domesticated, or rather, so kept; for it is doubtful whether the others, mullet, for example, also bred in captivity. As for fresh-water fish, the Romans were before us in raising young fish for stocking or re-stocking lakes and rivers; they also kept such fish in ponds, but how far this went towards domestication is not at all clear.

We should, I think, conclude that in this category muraena was the only fish that was completely domesticated. There is no evidence that the Romans ever bred and farmed carp, yet farmed carp were in use so early by their barbarian successors as masters of Europe that one is tempted to think that they must have done. The explanation may be found in the fact that, in the West, the carp was native to the Danube system of rivers and that some of the half-Romanized barbarians first began its domestication. It is to the point that King Theodoric the Ostrogoth insisted on carp for his table.

Until relatively modern times there was comparatively little immediate exchange, or contact of any kind, between the civilizations of western Europe and the Far East; so little that we must suppose that there were two quite independent domestications of carp, one in Europe and one in China. This fish began to be important in the menu of the upper class and the clergy from about the fifth century. Its importance as a farmed fish was promoted by the demand for fish created by the keeping of Friday and Lenten abstinence from flesh-meat. Supplies of fish not dependent on fishermen's luck and industry were therefore important. The farming of carp, in Roman-style stews, was carried on mainly by the monasteries, and that the carp really was domesticated is certain:

The carp has developed definite domestication characteristics. Apart from the normal grey coloration, red, white and
mottled carps occur. The cover of scales may be normal, or
there may be a few rows of very large scales, or the fish may
be devoid of scales altogether.[1]

It should be added that this was true both in the West and
in China; and in both those parts of the world the practice
of fattening carp by excessive feeding (in Europe on bread
and milk, the carp being removed from the water and kept
in wet moss) was developed.

The Crucian carp, or Goldfish, is likewise a completely
domesticated animal with many attributes and ornamental
deformities quite unknown in its wild state. It was domesticated by the Chinese, beginning probably about A.D. 960,
under the Sung Dynasty, the object being more food. But in
the course of breeding in captivity mutant fish with very
ornamental qualities occurred—not only in coloration, but
also in the shape of fins and tail. These were selected and
segregated as decorative fish; and as more and more fancy
fish were bred, fish-breeders became more interested in
producing valuable specimens for the rich aesthete than
ordinary ones for the fishmonger. There is, by the way, no
surer mark of a high civilization than that reversal of priori-
ties. The Japanese had their first goldfish from China and
were soon producing fancy strains of their own. From China
also, fancy goldfish reached Batavia, were carried thence to
St Helena as a staging post and from that island reached
Europe in the seventeenth century.[2]

So the domestication of goldfish began as an economic
measure and ended in a decorative minor art. As for the
fourth of Zeuner's list of truly domesticated fishes, the
Paradise Fish was, from the first, captured from rice-paddies
and kept in bowls and tanks for its beauty. Bred in captivity,
it became more and more fantastic in form and colours,

[1] Zeuner (1963).
[2] Pennant (1776).

but although truly domesticated it was never of economic value.

It is biologically feasible to rear marine fish in the warm water effluent of electrical generating stations. Plaice and sole have been raised to marketable size in half the time it takes in the natural environment, but there are many problems still to be solved before a commercial technique is available.[1]

Some years ago an experiment likely to have important consequences in increasing the world's supplies of edible protein was started, as a result of some laboratory successes in breeding and rearing sea fish, at the Hunterston nuclear power station in Ayrshire, and at a coal-fired power station on the shore of Carmarthen Bay. This was one start in fish-farming which, since it included breeding fish, must be considered a step towards domestication of some marine fishes.

Coastal power stations use millions of gallons of sea-water every day, for cooling and other processes. This is poured back into the sea at a temperature considerably higher than that of ordinary sea-water in average conditions. In warmer water many forms of marine life, from phyto-plankton and zooplankton to the higher fishes, grow faster and bigger. If you pass this warm water, on its way back into the sea, through suitably large open tanks or ponds so connected with the sea that any fish you put into them cannot get out, while predatory fish in the open sea cannot get in, then the fish in your tanks will enjoy an exceptionally favourable environment. Protected from predators, a higher proportion of young fish will survive than in the wild state; they will, as I have said, grow faster, especially since their feeding can be controlled.[2]

The fish in the power-station trials (carried on by the White Fish Authority, a Government agency) were plaice and sole. In both cases results were positive, the sole doing better

[1] Summary heading to Nash (1968).
[2] Nash (1968).

than the plaice, but both reaching marketable size about a year sooner than they would have done in natural conditions. As a result, trials were continued at Hunterston on a more important scale, the young fish to stock the new range of tanks being flown from a Ministry of Agriculture and Fisheries marine-fish hatchery in the Isle of Man. This hatchery, dealing chiefly in sole, was established as a result of the work of J. E. Shelbourne, who applied discoveries made in the late 1930s by Gunnar Rollefsen, notably that sole do very well on a diet of the shrimp (*Artemia salina*) provided from the hatchery's completely automated shrimp-egg incubator.

Results of these further experiments at Hunterston, although they would not yet have tempted a capitalist, were again positive. There was and is still a great deal to be learnt, but it is now clear that domestication of sole and plaice, and their farming on a large scale, are perfectly feasible. It is, as we have said, true that the keeping and farming of wild, or even of specially hatched, fish is not complete domestication; but can there be any doubt, in the light of the whole history of this subject, that selective breeding will ultimately follow?

By 1968 the director of the Hunterston project could write:

Although we do not know enough to make a rational design of tank systems, it is clear that power station discharges are suitable for sole and other marine species. What we must do now is to start selecting those other species which may be directly or indirectly useful in fish farming and to follow their tolerance and growth in trials like those originally performed at Carmarthen Bay and Hunterston. A considerable effort will then be needed in practical laboratories to produce culture techniques for the selected species. Fish, shellfish and possibly commercially needed algae may all prove possible to culture in tanks, ponds or open lagoons around a power station outfall. Stratifying or mixing populations may cut capital costs besides creating efficiency through commensalism. Moreover, since the warm waters of a coastal station provide temperature conditions found naturally in other parts of the world, there is a chance to try and culture and farm clams, ormers, large tropical prawns and many other shellfish and fish which, not insignificantly, command a high price. The technique for fast growing individuals may be no more than an operation of over-wintering in the first year, and the cost of installing filtration or chemical treatment units for a short period in the growth cycle and for resident spawning stock may not be prohibitive.

It is possible to see that even now the power station environment has economic significance. Future development must depend ultimately on the policies of the Electricity Boards producing the primary resource, the warmed water. Maybe we shall see private enterprise catering for a luxury market or, on a national basis, endeavouring to make more protein available for this country and for others, perhaps creating coastal complexes of power station and bay barrages to produce both marine and freshwater species. All these developments are distinctly possible but the speed at which they will be brought to fruition will depend a great deal on the amount of effort put into the science and technology of marine fish farming. Although it is growing, the effort is still not big enough or wide enough to meet the needs of those engaged in development and design. They need more and better knowledge about all aspects of the nutrition, health, environmental requirements and genetic possibilities of sea fish.[1]

A year later a Special Correspondent of *Nature* (Vol. 2, 24th October 1969) reported very favourably on the progress of this remarkable experiment, and concluded that it had been shown that farming of the sea is possible and could be an attractive alternative to hunting the sea, in the case of certain species:

There are exciting possibilities ahead. It may be that the most practical method of the future will be to rear fish to a certain size in warm water effluent at coastal power stations and then to transfer young fish to sea-farms in sheltered regions of the coast such as Scottish lochs.

What is envisaged here cannot, strictly, be called complete domestication. It can perhaps be compared to the first stage

[1] Nash (1968).

in rabbit-domestication when the animals were kept under control but living naturally. Yet, as I have said, it is difficult to believe that we shall not go on from there to selective breeding, segregation of mutants and so forth, just as the Chinese did with golden carp a thousand years ago and the European monks with carp in the sixth or seventh century.

But the kind of work which is being done at Hunterston (and also at Ardloe in Argyllshire) is not the only way in which late twentieth-century man is approaching the problem of domesticating marine fishes. The subject is so important that it will be worth while to glance at some others.

In the U.S.S.R. the approach has been different and, in my view, less advanced. Or perhaps it is fairer to say that the problem has been differently conceived. Soviet zoologists and botanists have long been interested in making use of the great natural history surveys at which they excel by finding ecological niches and fitting into them plants or animals useful to man. This, of course, is not domestication; it is a kind of assisting and guiding of nature. As regards fish, here is a case in point. Finding that the decline of whales in sub-arctic waters, due to over-hunting in the past, has left a surplus of natural food in those waters, Soviet scientists decided to introduce herring in large quantities, to live on that food. Another experiment has been the transfer of Pacific hump-backed salmon from the Pacific to the Atlantic; yet another and very successful one, the transfer of flounder from the Baltic to the Caspian.

The problem of supplying fish farms with young stock is being solved at the Isle of Man hatchery already mentioned. Some idea of the progress achieved is implicit in the following passage:

The [brine shrimp] eggs [for feeding sole and plaice], almost indistinguishable from sand grains, are imported by the plastic sackful from commercial producers on the salt flats near San Francisco. (When the San Francisco supply failed last year, a supply was obtained from the Great Salt Lake in Utah but for reasons not yet explained caused heavy losses among the larvae. Curiously, fresh-water larvae are not affected.) The eggs are loaded into a hopper from which they pass in measured amounts and at the appropriate intervals from one tank to another containing sea water at optimum temperature. This cycle continues without human intervention except to draw off the resulting food stock by the bucketful (looking rather like tomato soup).

The parent stock of plaice and soles dwell in large sea-water ponds, fed like fighting cocks on minced boiled mussels and 'queens' (a kind of small scallop). It is an odd sight to see a large plaice standing out of the water almost to its 'shoulders', in expectation of a between-meals titbit.

When spawning takes place (mid-February to the end of April) the eggs are gently skimmed from the surface of the pond by a large fine-mesh net and placed first in a transfer vessel and so to the hatching and rearing tanks of black polythene which stand in racks in a large room cooled to an air temperature of six degrees C. Incidentally experience suggests that it would probably be cheaper for the fish farmer to buy his stock at the egg stage, transferred—perhaps from a national stud—in large-mouth vacuum flasks which would take comparatively little room in an aircraft. One million sole eggs in the minimum amount of water weigh about three kg. On the other hand 50 to 100 small fish must travel in 3 litres of water.[1]

This part of the work is particularly important because, although there is nothing new in hatching out marine fish eggs, even in great quantity (the thing has been done here, in America and in Norway for more than half a century), the next stage, the feeding of larval fish, is new. Formerly the young fish had to be released into the sea at a very early stage, in the hope rather than the expectation that enough would survive to adulthood, so to speak, to make the work worth while by swelling the numbers of fish available to commercial fishermen. So that the work of J. E. Shelbourne, mentioned above, based on Gunnar Rollefsen's discovery of a suitable live food for larval fish of certain species, and

[1] Low (1967).

his trials at the Lowestoft laboratories, have completely transformed the prospects.

Although most of the work has been done with plaice (*Pleuronectes platessa*) and with sole (*Solea solea*), there are plans to use the same methods with turbot, lemon sole and, in due course, other species. Two different methods for dealing with the post-larval stage were considered. One is called 'low-density' culture, which means using large stretches of water in which natural productivity might be increased by the use of artificial fertilizers, the fish being contained by barriers. The other is 'high-density' culture, in which the fish are held in enclosures made of netting, in the open sea, or in shore-based tanks, and in any case fed with supplementary foods.[1] A combination of the British and Japanese methods, described below, here suggests itself; for the Japanese are experimenting with the 'herding' of fish by means of sonic-wave calls and electric field barriers.

The White Fish Authority has decided to concentrate its very limited funds on 'high density' experiments of the kind described above as being carried on at Hunterston and Ardloe. A number of difficult problems have still to be solved: at Ardloe, for example, otters, crabs and birds took a toll of the young fish, and there was the difficulty of maintaining salinity at the correct level after heavy rain. The best results have been obtained by the use of floating tanks for fish up to 8 cm. long, and thereafter of netting enclosures on the sea-bottom.

That the Authority has in mind not only the mere 'herding' of fish but also real domestication is clear from the following:

It is not essential to the success of fish cultivation that plaice should be hatched artificially since young plaice are plentiful on some parts of the UK coast and should be caught easily in quantities sufficient to meet the needs of a considerable

[1] White Fish Authority Report (1970).

number of farms. However this is not true for other species such as sole, turbot and brill. Moreover the ability to hatch fish offers several benefits: e.g., provision of a dependable supply of fish at a known cost, *and the possibility of selective breeding of new strains, hybrids etc.*, which are more suited to the conditions in which the fish will be reared, and which embody characteristics such as fast growth, good food conversion rates, resistance to disease, etc.

Further advantages arising from the ability to rear fish artificially are that the growth is not checked by transfer from the natural environment and the fish are easier to wean onto the non-living foods which are used to grow the fish to market size. An apparent disadvantage, which may in practice not prove to be significant, and indeed could be an advantage, is that a certain proportion of the artificially hatched fish are poorly pigmented and can be completely or partially albino.

Because plaice and sole can be hatched in quantity, these species and other high-cost species such as turbot and brill are being used at the Authority's sites to develop cultivation techniques. The species which will eventually be found to be most suitable to commercial farms are not known. The position is similar to that in which the farmer found himself many centuries ago; he did not know which animals would be domesticated and be the sheep, cows and the poultry farmed today.

Ultimately a marine fish farm is likely to use a variety of species which can be grown together to make maximum use of the volumes of water available; flatfish on the bottom, round fish swimming above, and perhaps crustaceans and molluscs will figure in this mixed culture . . .[1]

If we consider the case of the vertebrate fishes, Britain, thanks to the White Fish Authority, is far ahead of the rest of the world in this field, the farming and ultimately the domestication of marine fish. Other peoples have worked in different fields. I have already mentioned the Russians, and turn now to the Japanese.

In Japan marine farming has started on quite different lines. There, commercial exploitation of shrimp and prawn farming is far advanced. At Ikushina a firm called the

[1] White Fish Authority Report (1970).

Shrimp Farming Co. has a prawn-egg hatchery where prawns
are reared until they are between $1\frac{1}{2}$ and 2 cm. long. At that
stage they are sold to the prawn farmers of the Seto Inland-
Sea Marine Development Co., who grow them on to marketable size and sell them. These specially fed farmed prawns
are said to be of superior flavour. There are now some other
firms engaged in the same industry. In our context an interesting aspect of their work is the way in which the eggs are
produced at the hatchery: the farm buys mature prawns
from local fishermen and pairs are put into tanks, several
pairs to each tank. The prawns spawn between 300,000 and
1,000,000 eggs per pair, and of these 20 per cent (an enormously higher proportion than in the wild state) survive to
the 2 cm. stage.[1] This single hatchery at Ikushina is producing about 50,000,000 prawns a year. But let us take into
account the possibility of the prawn-farmers going a step
further, selecting parents with superior attributes, isolating
them, and then studying their offspring for valuable modifications. Obviously the difficulties are great: for one thing,
they entail looking at between 60,000 and 250,000 2-cm.
prawns at each hatch. Yet I have every confidence that it
will in due course be done, first for size, by screening, and
later, for other attributes, perhaps by electronic scanning.
For the very method used at the hatchery makes possible a
considerable measure of control, once market pressure makes
it worth while to begin the kind of improvements which are
typical of domestication.

The Japanese method is obviously applicable to other
crustacea, for example, scampi, crawfish and lobsters.

The Japanese have not confined themselves to work with
crustacea. In 1969 the government fisheries agency decided
to begin fish farming on the continental shelf: 'A limited
area will be equipped with food-boxes fitted with timing
devices to control feeding intervals and fish will be attracted
by sonic wave signals reinforced close to the food-boxes by

[1] Low (1967).

32 and 33. The way in which certain sub-arctic peoples live parasitically upon reindeer herds teaches us how other kinds of grazing and browsing animals were probably domesticated in the remote past. Nomads can move with the herds, killing for meat as they need it. Semi-settlement entails controlling the herd movement, out of which develops taming of the animals for draught and burden, and finally controlled breeding in captivity.

34

34. The Arabian camel, *Camelius dromedarius*, is almost certainly extinct in the wild. It was domesticated much later than the horse, which probably suggested the possibility of domestication by capturing young camels and rearing them in captivity, to work in steppe conditions impossible for horses, such as deserts.

35

35. *Camelius bactreanus* was domesticated in Central Asia to work in conditions which were too exacting for horses or even asses. Broadly speaking, *C. dromedarius* has been modified by selection and breeding for speed as a mount, and can cover up to ninety miles in a day; whereas *C. bactreanus* has been adapted for endurance, carrying heavy burdens up to twenty-five miles at a stage. *C. bactreanus* is believed to be not quite extinct in Central Asia, but the small herds observed may be feral animals. Feral dromedaries are numerous in Iran.

36

36. The Guanaco of the Andes high plateaux belongs to the *Camelidae*. The great herds were first hunted by pre-Inca peoples, later domesticated as the *llama*, for use as pack-animals and for their wool, meat being provided chiefly by the hunting of the wild guanacos.

37. The Vicuña of the Andes, a smaller relative of the llama and the alpaca, is a case of domestication prematurely abandoned. Its fleece yields the finest wool of all; the pre-Inca and Inca peoples hunted but did not domesticate it. First successful domestication was accomplished by the Jesuits at their missions in the eighteenth century but their work was abandoned when they were expelled and has never been resumed. Probably in danger of extinction.

38

39

38. Elephants were first domesticated for work and war in India,
but the notion that African elephants cannot be tamed is nonsense.

39. Indian elephant with, on the left, an African calf and, on the right, an Indian calf. Note the difference in the ears. Fully grown, the African elephant is the larger species.

40. Domesticated Indian war-elephants presented as a state gift led the Carthaginians to undertake domestication of the African elephant and its use in war against the Romans. War elephants captured in Alexander's Indian campaign led to their use by the Greek kings of Successor States.

41. The difficulties, long insuperable, of domesticating the South American fur-bearing Chinchilla had nearly led to its extinction in the wild, but they were overcome by an American engineer and the species is now being modified by selective breeding.

42. A case of twentieth-century domestication, the Mink's fur colour has been modified by selective breeding of mutants to yield the 'blue' and 'palomina' furs of commerce.

41

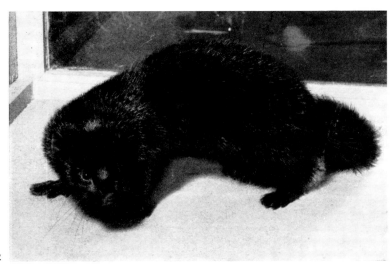

electrical shock barriers to shepherd the fish directly to the food.'[1]

This experiment is founded on the invention of a superior fish-food based on one of the yeasts. The Japanese have an advantage in their long experience in fresh-water fish farming; and one of their fresh-water fish, the rainbow trout, is quite happy in sea-water and is being used in this work. It is expected that as well as fish, lobsters will be farmed in this manner, a kind of wild fish-herding by electrical devices. Japanese scientific workers in this field have calculated that if only 10 per cent of their continental shelf is used for fish farming, their national shortage of animal protein will be made good. This would mean establishing fish farms on 26,000 square kilometres of continental shelf, so that there is still a long way to go.

The future of the domestication of marine fish is being prepared for by studies of the feeding habits of fish, and of their methods of finding the food they prefer. At the Bedford Institute's Marine Ecology Laboratory, Dartford, Nova Scotia, a thorough study of the feeding habits of an important food-fish, cod, has been and is being carried on. The results of such work will enable fish farmers to feed their fish on compounds which have the qualities of the natural foods but which are cheap, and to know how to feed for optimum growth and growth rate.[2] Of the small bottom-dwelling animals on which cod feed, why do they prefer one species to another? How far will they go to find a particular food rather than eat what is handy? Is a preferred food necessarily the one on which they thrive best? These are the kind of questions which have to be answered before we can undertake the domestication of so wide-ranging a fish as the cod.

[1] *New Scientist*, 6th March 1969, p. 496.
[2] Brawn (1969).

MOLLUSCS

Molluscs are not fish, but as all but one of the commonly eaten ones are marine animals, they will not be out of place in this chapter; and as the only land mollusc which is commonly eaten, the snail, is hardly worth a chapter to itself, I will include it also here.

Molluscs formed an extremely important part of the diet of Palaeolithic man in many widely separated parts of the world. Some communities subsisted almost entirely on molluscs, which are very nearly a 'whole' food; in doing so they accumulated, for example over an immense part of coastal South America, great mounds or middens of shells which have long engaged the attention of archaeologists and palaeontologists. Nor has the eating and the industrial use of molluscs ever ceased: for hundreds of thousands of years proto-man and man have been gathering shell-fish, eating the animal and using the shells, or some of them, as tools, weapons, ornaments and raw materials for industrial processes and works of art. The two oldest economic acts we still perform are the gathering of wild fruits, such as blackberries and cranberries, and the gathering of mussels on the seashore.

Two mollusc genera, *Murex* and *Nucella*, enriched the Phoenicians and were of importance to the Minoans. From a gland in the shells came the dye called Tyrian Purple which was so fashionable 3,000 or 4,000 years ago and was still important in Roman times. In his Ninth Book, Pliny gives an account (Philemon Holland's translation) *Of the greatest Winkle called Murex*:

But that beautiful colour, so much in request for the dying of fine cloth, the Purples (Murices) have in the midst of the neck and jaws. And nothing else it is but a little thin liquor within a white vein; and that it is which maketh that rich, fresh and bright colour of deep red purple roses. As for the rest of this fish, it yieldeth nothing.

Pliny goes on to tell how the dye was made. His instructions have been of no use to later workers trying to rediscover a secret lost in the twelfth century, but one suspects that it would very soon be recovered if there were any call for Tyrian Purple today. As each shell had to be struck at a specific point to detach the 'vein' containing the dye, and as 8,000 shells were required to produce one gramme of dye, it is not surprising that Tyrian Purple was dear and made the fortunes of its purveyors though not of its producers who were, of course, slaves.

That was only one of the industrial uses to which ancient man put mollusc shells. Much earlier was the use of certain shells as tools, like the Polynesian adze-blades made from *Tridacna* shells and used in boat-building. All round the Mediterranean, Neolithic man made use of the cockle, *Cardium*, to impress patterns on pottery, and later the cardium pattern was imitated with a graving tool which might be the potter's finger-nail. Mother of pearl, the nacre with which oyster and other bivalve shells are lined, was already being used by craftsmen at Ur of the Chaldees; and some of the grave furniture of the Pharaoh Tutankhamen was inlaid with nacre. In southern Asia and in parts of the southern hemisphere and Africa the shells of the cowrie, which also have sexual significance in a number of religions, were so widely used as currency that when this gasteropod received its Latin name it was called *Cypraea moneta*. Mussel shells also were used as currency in some parts of prehistoric Europe; and pearls, both of oysters and mussels, were probably the earliest precious gems used by man.[1]

One would have expected all this to have led to some kind of domestication, were it not that the difficulties of real domestication are very great indeed. On the other hand, there is no difficulty in *farming* marine molluscs; and yet even that was not, as far as we know, attempted until fairly recent times.

[1] Evans (1969).

Palaeolithic men seem not always to have eaten molluscs from choice. They were driven to that diet by environmental changes. Evans says that in western Europe the adoption of mollusc-gathering as a way of life is associated with the spread of forests after the last glaciation and a consequent restriction of open areas available to Mesolithic man for hunting. In California (Canaliño Culture), on the other hand, it was associated with a decline in the amount of game following a reduction of forest area by drought. Still other causes drove Mesolithic man in other places to adopt a diet of molluscs, which, although sufficiently nourishing, demand a lot of labour for a small return. However:

... while exploitation may lead to changes in the composition of shell-fish populations there is no evidence to suggest that shell-fish have ever been domesticated; i.e. selectively bred to produce a stock which differs genetically from the wild ancestor. Domestication as a natural process under conditions of association with man is highly unlikely when one considers how remote is the biological contact between man and mollusc, each occupying an entirely different life medium. And exploitation will never lead towards domestication, only away from it sometimes resulting even in extinction. Deliberate domestication, too, is a practical impossibility since the method of reproduction involves the release of gametes into the sea, so that there can be little control over mating.[1]

The nearest we have yet come to the domestication of molluscs is oyster and mussel farming. A number of countries have oyster and mussel farms, but in this field the French are in the lead. Such farming does not amount to much more than putting seed-oysters into new water. Still, I cannot see why Evans makes quite such a point of the difficulties in the way of true domestication; would not the gametes be released in closed tanks? Even in the open sea some results might be accomplished by, for example, mixing Japanese and New Zealand oysters with our own. Those oysters are two more of the mollusc kind which are now being farmed.

[1] Evans (1969).

Others again are the objects of work tending to convert them into sea-farm animals; among these are the big Chilean mussels and the American clams (*Venus mercenaria*). There is a reason for our special interest in these clams, for by a curious accident they have become naturalized in British waters. A colony of *Venus mercenaria* was found established under the warm-water outfall of the Marchwood power station on Southampton Water. It seems likely that this has been brought about by the practice of throwing overboard, from transatlantic liners, the remains of the ship's store of live clams carried for the dining-rooms on the west-to-east passage. But as the clams do not breed with regularity, a pilot project was inaugurated, to try to rear young clams in a hatchery; and there is at least a prospect of large clam-farms in our southern estuaries.

SNAILS

The existence in North Africa of what French scientists call *escargotières*, large middens of snail-shells accumulated by men of the Capsian culture, is good evidence that some pre-historic peoples lived in part and perhaps chiefly on these land molluscs. There is even some evidence for what one might call a tentative and unconscious move towards domes-tication of one species in the fact that in Iran *Helix salomonica* was eaten in much larger quantities than other species equally succulent and wholesome, and just as abundant.[1]

The Romans, however, were the first people actually to domesticate an edible snail, *Helix pomatia*. According to the chronicler Sahagun, the Aztec menu included either snails or slugs, elaborately dressed and cooked. I have not been able to discover whether they were domesticated, probably they were just collected in the wild; but as they were served in the royal palace, where over a thousand people fed every day twice a day, it is at least possible that the snails and slugs

[1] Reed & Braidwood (1960).

were 'farmed' in some kind of enclosures. The Roman snails, however, were truly domesticated animals, in that potentially superior variants were selected, paired in vivaria for breeding, and strains remarkable for size, colour or fecundity segregated. Evans (1969) says that this was more of an academic experiment than a policy dictated by economic hardship. He here misses the point: there was no need of economic hardship, of course, to persuade the Romans to try curious kinds of food, such as 'Fat' dormice produced with great trouble and expense. The domestication of the edible snail was dictated by Roman *gourmandise*.

Since such edible snail species as *Helix aspera* and *H. pomatia* have, as a result of environmental changes brought about chiefly by the spread of man's civilization, become entirely dependent on man's gardens and plantations for their survival,[1] it would be particularly easy to resume domestication of the snail and farm it on a considerable scale in many parts of the world. But at the time of writing only the French and Belgians seem interested in edible snails; they still, as the Romans did, farm them in snail-gardens—the Roman *cochlearia*. I shall have a little more to say about edible gasteropods in the final chapter.

[1] Kerney (1965).

14

Omissions and Conclusions

A majority of the animals discussed in this book, and certainly the most important ones, were domesticated in pre-historic times, as were all the most valuable of the economic plants discussed in my previous volume.[1]

In order to understand why this was so, and why, since the beginning of History proper, that is since the invention of writing, animals never domesticated by prehistoric man have been neglected by historic man and this kind of economic activity fell away to almost nothing, one must begin by rephrasing the opening statement of this chapter. For that statement implies that 'History' was a pre-ordained and inevitable phenomenon, meaning by 'History' not the account but the fact of the rise of ever more wealthy and complex civilizations based on ever more complex ideas and technologies. But there was nothing pre-ordained about it; man could have remained no more than a species of mammal occupying a niche in the world's ecological fabric. Man, as Gordon Childe long since pointed out, made himself. He began with the discovery of mind and will.

So the opening statement ought to read something like this: 'History' began, that is man was set upon an ascending course, because it happened to him to domesticate certain plants and animals, to submit himself to their service that they might return him service a thousandfold.

There was an Age of Domestications; as a result of his activities during that Age, man found himself rich; and being rich he could give expression to the idea and the will he had found in himself, or which had grown in him as a result of his

[1] Hyams (1971).

own ever more subtly exercised ingenuity. Man made his
mind and soul as an athlete makes his muscles, by use; each
use suggested other uses; and successful use, as in the
domestications we have discussed in this book, gave him
the vision to see and the means to go further.

But it has repeatedly happened in this story that when a
new idea, expressed in the beginnings of a new technique,
took possession of men's minds, the older idea in the same
field of work was prematurely abandoned, that is, abandoned
before the utmost had been made of it. I do not suppose that
when stone tools were displaced by metal they had been
brought to a point of perfection which could not be improved
on, even though metal and stone tools co-existed for thousands of years. To make a more modern example: at the
very moment when the design and rigging of sailing ships
reached a point of excellence, and a whole new era of progress in that field seemed probable, steam began to replace
sail and the attention of shipbuilders was distracted from
what might have been achieved. Even more important is the
case of the motor vehicle; the invention of the internal com-
bustion engine turned the attention of engineers away from
work on steam-engines for road transport and the use of
electric traction; as a result we are stuck, for the time being,
with a device which is noisy, dirty, and at its very best
extremely inefficient.

The great Age of Domestication having, by enriching us,
given us liberty to turn our attention to other business, we
never, as it were, quite finished the older job.

I have already referred briefly to certain ungulates whose
domestication was never completed and which are now
either very rare in domestication, or survive only as wild
animals. It could be said that not one of them could do for
us anything which is not done better by one of the animals
we already have in domestication. But this is not quite true.
Some of them would enable us to exploit kinds of pasture
and kinds of climate where they would thrive and our own

cattle would not; in a world where more than half the population suffers from a shortage of animal protein, we have no
right to neglect any such source unless we find ways, very
quickly, to stop living on animals and live entirely on vege-
tables, which would, by the way, be more economical of land.
Then there is the desirability of variety. Where is the house-
wife or caterer who has not, time and again, complained that
the eternal round of beef, mutton and pork becomes a bore?
We might have had half a dozen more kinds of meat to choose
from, half a dozen more kinds of hides to use.

We no longer, it is true, have any use for fast-running elk
to carry the mails, either as mounts or draught animals; nor
has their milk any advantages that I know of over cow's
milk, or their flesh over reindeer meat. But perhaps there
may be niches in the 'domestic animal ecology', if such a
phrase is permissible, into which gazelle, antelope of various
kinds, ibex and bison could usefully be fitted. The very fact
that this consideration seems academic to the point of futility
proves my point that we are, so to speak, no longer really in
the mood for domesticating animals.

There cannot be much doubt not only that we could complete the Jesuits' work of domesticating the vicuña whose
hair yields after processing a cloth of incomparable quality,
but also that a domesticated vicuña could be used to bring
prosperity to poor mountaineers in all the great mountain
ranges of the earth, and not just the Andes. It is not good
enough to dismiss this by saying that we now make our yarn
in factories; the natural yarns are still better, and mixtures of
natural and artificial fibres best of all. And even if total
reliance on man-made yarns were possible, why should we
be less rich in a variety of fabrics than we need be?

There are some interesting omissions from the list of
animals which have been domesticated, which are worth a
few minutes' attention.

First there is, strikingly, the whole order of the Reptiles.
At first sight it might seem that there would be no point in

domesticating any of them, but it would not be quite true; and, when thinking on these lines, one must try to imagine the wants and lines of thought of the ancients, rather than our own.

Take first of all the snakes. In many parts of the world snakes have been and are eaten; by some peoples they are relished. The non-venomous ones are harmless; they are not at all difficult to tame and could easily be 'farmed'. In some parts of India house-snakes were kept, fed, petted, their work being to keep the house free of mice and other small vermin; they were not, however, domesticated, they were simply tamed. In our own time there are snake farms in America, Africa and Australia, for the making of anti-venins based on snake-venom; but again, the snakes are not domesticated, merely kept in captivity. Much use is made of snake-skin in the manufacture of shoes, handbags and other fashion accessories; yet nobody farms snakes for their skins, much less undertakes a programme of controlled breeding to produce superior skins with new colours and patterns. Aesthetically, as well as economically, domestication of some snakes might prove most rewarding; and precisely the same is true of lizards, which would be even easier to handle and whose skins are even more valuable.

FROGS

Frogs (belonging to the Order Amphibia) have been eaten by man in many parts of the world since prehistoric times and have become gastronomically important in the cuisines of two great cultures, the Aztec and the French. In Europe the edible species are *Rana esculenta* and *R. catesbiana*; in other continents half a dozen different species are eaten. The frogs served to the Aztec nobility in large numbers were caught in the wild; so are the frogs eaten in France. Yet frogs would be very easy to domesticate and improve, by the isolation of selected pairs in tanks.

It is true that, as food animals for man, reptiles and amphibians have been and remain only marginally important. The world trade in edible frogs does not amount to $10 million. As a rule such creatures have been eaten by two categories of men: the very primitive and the ultra-sophisticated. The ordinary middle kind of people, the great majority of mankind, have never had much taste for eating snakes, frogs or lizards, although all three have a delicate flavour and texture. There is another thing against snakes and lizards which has saved them from being devoured by man, and from domestication: although the majority of reptiles are harmless and inoffensive animals, the whole Order has, in the popular imagination, been condemned as venomous by association with the few that are. Finally, there is an ancient and never fully explained antipathy between men and reptiles, which might almost seem to be an inherited memory passed on to man by some remote pre-hominid ancestor of the days when the great reptiles ruled the earth.

TURTLES AND CROCODILES

Turtles and tortoises, on the other hand, are not offensive to man and are not popularly thought of as reptiles at all. The meat of the turtle is considered a luxury food by many civilized peoples; the eggs are relished by many less civilized; and tortoise-shell is an ancient material for the craftsman. Slow, tractable and of very manageable size, many species would be easy and profitable to domesticate, yet to the best of my knowledge the thing has never been attempted. In the mid fourteenth-century *Voiage and Travaile of Sir John Mandeville*, its author claims that certain peoples ride on the backs of giant sea-turtles. His entire book was later dismissed as a pack of plagiarisms and lies; most of it was, but as it happens he was telling the truth about turtle-riding. There is, however, something too fantastic about the idea to allow us to think that these great marine turtles might have been domesticated as mounts.

Crocodiles and alligators have frequently been tamed, and alligators are now farmed for their skins. But they are simply kept in captivity, and no attempt has yet been made to domesticate them. There might seem to be no point in trying to do so, yet it is an undeniable fact that the size, quality and colour of skin could be much improved by selective breeding and that the relentless growth of man's numbers will increasingly reduce the amount of territory available to the wild crocodiles.

The argument that the biological remoteness of man from the reptiles makes domestication impracticable will not really do. Man has domesticated creatures as remote from him biologically as bees and moths and fish. Most of the reptiles are at least, like ourselves, land animals.

MORE GASTEROPODS

A number of gasteropods were mentioned in the last chapter, and the fact that there is no serious difficulty in domesticating snails has been referred to. In addition to the edible snails, which have, in the past, been domesticated, there are at least two other snails which, although their economic importance is much greater than that of the Helix species, have not been domesticated but certainly could be.

First there is the big marine snail already mentioned, the 'ormer', *Haliotis abalone*. This animal lives on rocks and seaweed on certain Californian and Australian shores. It has a large and muscular foot which, cut into slices and pounded to make the meat tender, is usually cooked by frying in bread-crumbs, like *escalope* of veal. This food is sufficiently well liked—that is, there is sufficient demand for it—to make canning worth while; and, like the flesh of nearly all molluscs, it is wholesome and nourishing. This big gasteropod was eaten by some coastal tribes of American Indians, and from them the Spaniards acquired their taste for it. It could easily be farmed in closed sea-water lagoons and might as easily be improved by selective breeding as was *Helix pomatia* by the Romans.

The other big snail which could be, but never has been, domesticated, although it is a valuable and much sought after food for man, is a land animal confined to the tropics, *Achatina*. This splendid beast, with a shell six inches in diameter, is eaten by some African peoples, and probably has been since man's hominid ancestor got his living by seeking grubs and insects. This again could as easily be farmed and ultimately domesticated as the Helix species.

There are some other omissions. No marine or aquatic mammal has ever been domesticated, neither otters nor beavers, nor any of the whales, seals or sea-lions. Perhaps there is no obvious reason why we should ever have undertaken their domestication. All kinds of whales are easily tamed if captured young enough to be of manageable size during the difficult early stages. Dolphins are said to be the most intelligent of all animals; so intelligent, indeed, that the most absurd claims have been made for them, as that they can be taught to speak. They might have been domesticated as a kind of sea-horses by a maritime people, but alas they never were, and the recent American suggestion that they be trained as a kind of animal *kamikaze*, to carry a charge of explosive against enemy ships, is one of the most repulsive ever made in the long story of the servitude of the beasts. Seals are famous for their tractability and intelligence, as also for their curiosity and pleasure in applause, all qualities which would make them easy to domesticate; and both their fat and their skins are very valuable to man. If the idea of marine herds of seals, controlled, as they could be, by electric barriers, as the Japanese are 'herding' certain fish, seems repellent when one considers that the animals would live only to be slaughtered in their prime, it is at least not so repulsive as the present methods of culling wild herds of seal. But nobody has ever tried to domesticate these attractive and easily tamed animals, and I do not expect that anyone ever will.

OSTRICHES, CANARIES AND FALCONS

Three kinds of bird-taming, two of which amount to domestication in full, have yet to be mentioned.

The ostrich has been of great interest to African man since Palaeolithic times in that continent. Later it was still hunted, and its eggs, collected by hunters, became valuable articles of trade, especially when painted, gilded or otherwise decorated. There was no ancient attempt at domestication. The Romans, in their jolly way, derived much pleasure from exhibitions of ostriches in the Circus, cutting off their heads so that the mob and the Senators could enjoy the spectacle of the big birds running about headless for some minutes. In Egypt ostriches were occasionally caught young and broken to harness: eight such birds drew the ceremonial chariot of King Ptolemy Philadelphus, whose queen sometimes rode an ostrich. But the Egyptians made no attempt to breed the ostrich in captivity.

In the nineteenth century, however, the demand for ostrich feathers led to domestication of these birds. It began about 1856 when the French started farming and breeding them in captivity in Algeria. This kind of farming was later abandoned in North Africa, but meanwhile it had spread to the south, where it is still carried on, although not on any considerable scale. Ostrich skin makes a very fine and delicate leather, which would ensure high profits from the exploitation of the ostrich in Africa, of the emu and cassowary in Australia and of the rhea in South America.

SONG-BIRDS

A number of species of song-birds, decorative birds such as budgerigars, and entertaining birds such as mynahs and

parrots have been domesticated more or less. In the budgerigar the process has gone far, with the production of
entirely new colours; it is, among domesticated birds, what
the Paradise Fish is among domestic fishes. But prototypical
is the canary-bird, and it will be sufficient in this category
to say a little about its domestication, for there are some
curious features in the history.

The bird is a finch, greenish or yellowish brown in nature,
native to the Canary Islands and bearing the scientific name
Serinus serinus canariensis. Because the naturalist Konrad
Gesner mentioned the domesticated bird in 1555, and because
the conquest of the Canary Islands was not completed until
the end of the fifteenth century (Grand Canary 1483, Tene-
rife 1493), Zeuner (1963) suggested that the bird may first
have been domesticated by the Guanches, the last remnant
of the great Cro-Magnon race, who were the Canarian abori-
ginals, and whose courage, Stone Age slings and primitive
Neolithic clubs gave the Spaniards, in their steel armour and
with their steel weapons, their cross-bows and their horses,
so much trouble. I find this improbable: it is clear from their
artefacts that the Guanches were at an early Neolithic stage
when the islands were conquered, and the only reason for
domesticating canaries is for their song. On the other hand,
there is a very singular reason which might explain a particular interest in bird-song: the Guanches were either the
inventors or heirs of a very remarkable practice of whistling
instead of speaking, in order to make their speech carry much
farther than shouted words will travel clearly. At the one
battle in which they totally routed a Spanish army, the movements of the various bodies of troops were controlled by this
means, a fact which goes far to explain the victory. The
Spaniards later borrowed this technique from the Guanches
and Spanish was whistled, as well as spoken, in the islands;
oddly enough, it is the only European language susceptible
of being 'sounded' in this way. The whistled Spanish (not
Guanche, alas) still survives, and is known as *silbo*, on the

island of Gomera. Now a people using such an odd technique of speech-projection might well have a very special interest in bird-song. Against this, however, there is another consideration which Zeuner did not, for once, take into account. I have seen many wild canaries on the islands of Lanzarote and Fuerteventura, both of which were completely conquered by the adventurer Jean de Bethencourt, by 1404, and settled with several hundred Norman families by 1406.[1] The Norman-Canarians, soon joined by many Italian colonists, had therefore a century and a half in which to domesticate the canary before Gesner first wrote about this bird.

Be that as it may, the Spaniards had for some time a monopoly of domesticated canaries, which they lost only when a ship carrying, among other things, a cage full of breeding canaries was cast away on the island of Elba, and the birds escaped to become feral and populated the island. The Italians seized this chance to become the principal purveyors of canaries bred as domesticated birds, and did very well out of the trade for more than a hundred years. By the eighteenth century, however, the town of Imst in the Tyrol had become the world centre of canary-breeding. By then pure yellow, primrose and even orange had replaced the greenish brown of the wild bird. In the nineteenth century the centre for canary culture shifted to one or two towns in the Harz Mountains.

BIRDS OF PREY

The birds of prey—including little sparrow-hawks, peregrines, gerfalcons, goshawks, eagles and owls, which have been used by man to help in hunting game-birds and small animals, even antelopes—have never been domesticated, for they do not normally breed in captivity. Very recently some

[1] Major (1872).

attempts to breed sparrow-hawks, in order to supply fruit-growers and viticulturists with a means of keeping fruit-eating birds off their crops, have shown promise. But since the young hawks established in an orchard or vineyard, though they remain for that season, are then uncontrollable, domestication cannot follow. Hawks can be tamed and trained, but no more.

The first people known to have 'manned' hawks for hunting were the Assyrians of Ashurbanipal's day; and the falconer's glove, the hood and the jesses were also invented in that part of the world. The Thracians were falconers, but the Greeks never took to hawking, nor did the Romans. But the Arabs, and in general the peoples of Asia Minor, Iran and Palestine, were enthusiastic falconers; it was from these Saracen enemies that the European Crusaders learnt falconry, acquired their first hawks and introduced the sport into Europe. Its greatest exponent was that emperor of genius Frederick II von Hohenstaufen, known as *Stupor mundi*, whose book *De arte verandi cum avibus* (Of the art of hunting with birds *c.* 1250), is not only a treatise on falconry which is still being reprinted, but also the first major work of scientific ornithology based upon observation and not on mere myth and legend.

But falcons of all kinds now, as in Frederick's day, are first taken in the wild and then 'manned'; they are not bred for their work.

There is, however, one bird of prey, a diver and fisher, which is truly domesticated: the Cormorant, *Phalacrocorax carbo*. These divers made their first appearance in England as domesticated birds, trained to fish for their masters, during the reign of James I. The royal cormorants probably came from Flanders, whence also the French Court had its first trained cormorants. Zeuner (1963) says that since it was from the Spanish Netherlands that Jesuit missionaries were sent to China, it was no doubt the Jesuits who introduced trained cormorants into Europe. For although cormorants are native to the Mediterranean and the Atlantic shores of Europe, they were never domesticated in the West.

The earliest accounts of domesticated cormorants are Japanese (fifth century A.D.). In Japan, no doubt, the domestication was accomplished, and from there domesticated cormorants and the technique of using them were introduced into China by way of Korea some time in the seventh century. The art has not been lost, and fishing with cormorants is still practised in Japan and in the Lower Yangtse Basin.

All civilizations were founded upon the cultivation of plants and the servitude of animals. It is probable that the latter, though not the former, could have been dispensed with. The Central American peoples created a great urban civilization, culminating in the Aztec Empire, without any beast of draught or burden, or any 'cattle' (except the turkey), or any wool-bearing animals. They prove that civilization can be erected on a foundation of domesticated plants alone. But the progress towards a machine technology of the Old World civilizations, and their breadth of view, would have been attained very much more slowly without the help of domesticated animals. Moreover, those societies would have been different: less inclined towards freedom of the individual but, by the same token, more stable.

The first close domestic tie formed between man and another animal was with the dog. By first rendering his hunting more effective, and thereafter by helping him to control the movement of wild herds of game and so taking the first step towards domestication, the dog hastened man's progress towards an economically superior way of life. No two higher vertebrates have ever formed so marvellously successful and profitable a partnership. The dog has ever since been rewarded for his help with board, lodging and love; and man, for his initiative, with the dog's utter devotion.

The control of herds of animals which could be lived on,

provided they could be given sufficient grass pasture, made possible the accumulation of wealth without having recourse to the farming of plants. The men who served, in order to be served by, animals retained a measure of freedom of movement, and therefore of the spirit, which was enhanced when they also learnt to ride on animals. This was impossible to those who served, in order to be served by, plants. The pastoralist everywhere developed qualities of mind and spirit different from those developed by the sedentary peasants. It was, on the whole, among the peasant peoples that such institutions as slavery first developed; but also the habit of careful observation of nature leading to an ever deepening understanding of her laws, that is to science and philosophy. It was, on the whole, among the herdsmen that the qualities of the individual man emerged as important, and therefore the idea of freedom. The herdsman's gods were sun and sky and the male principle; the peasant's, earth and water and the female principle.

When the pastoralist way of life conferred on its practitioners freedom of movement over vast territories, a pattern of contact followed by association was formed between those two kinds of peoples, and repeated itself all over the world. The herdsmen—Aryans from Central Asia, descending upon the peasants of India and of Europe; the Semites of the Arabian peninsula and its margins descending upon the peasants of Mesopotamia and Palestine like the Israelites on Canaan; the Mongols and Tartars of Asia striking both east and west to overwhelm the sedentary societies of China and Russia; the Inca and other Andean mountaineers pouncing on the coastal peasant cultures of the western watershed of the Andes—seized upon the farmers' civilizations and, in time, were absorbed by them. They taught their victims their skills and ideas and gods; their victims did as much for them. From that mutual teaching came syncretic religions, cultures, philosophies, sciences.

Out of those fusions rose societies richer in ideas as well

as things, based on cities whose increasing wealth was created by both plant and animal husbandry. From those societies came the greatest civilizations of mankind.

The domestication of plants and animals was not a product of civilization: it was its origin.

Bibliography

AITKIN, F. C. (1963). *Feeding Fur-bearing Animals*. Farnham Royal.

ANTONIUS, O. (1918). *Stammesgeschichte der Haustiere*.

ARAMBOURG, C. (1968). 'Les Données de la Palaeontologie humaine.' In Varagnac (1968).

BARNETT, R. D. (1939). 'Phoenician and Syrian Ivory Carving.' *Palestine Exploration Quarterly*, 4. London.

BERRY, R. J. (1969). 'The genetical implications of domestication in animals.' In Ucko & Dimbley (1969).

BOSCH-GIMPERA, P. (1968). 'L'Amérique Paleolithique inférieur de l'Europe et de l'Afrique.' In Varagnac (1968).

BRAWN, V. M. (1969). 'What's on the codfish's menu today?' *New Scientist*, 6th March 1969. London.

BREUIL, ABBÉ H. (1968). 'L'Art Palaeolithique.' In Varagnac (1968).

BREUIL, ABBÉ H., & LANTIER, R. (1959). *Les Hommes de la Pierre Ancienne*. Paris.

BUSE, H. (1969). In *El Commercio*, 31st October 1969. Lima.

BUSHNELL, G. H. S. (1957). *Peru* (Ancient Peoples and Places series). Thames & Hudson, London.

CHILDE, G. (1951). 'The First Wagons and Carts from the Tigris to the Severn.' *Proceedings of the Prehistoric Society* (N.S.), 17 (3).

CLARK, G., & PIGGOTT, S. (1965). *Prehistoric Societies*. Hutchinson, London.

CLARKE, J. D. W. (1961). *Modern Chinchilla Farming*. Toronto.

CONWAY, M. D. (1879). *Demonology and Devil-lore.*

COON, C. S. (1951). *Cave Exploration in Iran, 1949.* Pennsylvania University Museum monograph.

CURASSON, G. (1947). *Le Chameau et ses maladies.* Paris.

EVANS, J. G. (1969). 'The Exploitation of Molluscs.' In Ucko & Dimbleby (1969).

FLANNERY, K. V. (1969). 'Origins and ecological effects of early domestication in Iran and the Near East.' In Ucko & Dimbleby (1969).

FRANKFORT, H. (1939). *Cylinder Seals.* Macmillan, London.

FRASER, H. M. (1951). *Beekeeping in Antiquity.* London University Press.

GORDON, SIR T. E. (1906). *A Varied Life.* London.

GOWERS, W., & SCULLARD, H. H. (1950). 'Hannibal's Elephants Again,' *Numismatic Chronicle* (6), 10 (39–40).

GRIGSON, CAROLINE (1969). 'The uses and limitations of differences in absolute size in the distinction between the bones of aurochs (*Bos primigenius*) and domestic cattle (*Bos taurus*)'. In Ucko & Dimbleby (1969).

HATT, G. (1919). 'Notes on reindeer nomadism.' *Memoirs of the American Anthropological Association,* 6.

HAUK, E. (1950). 'Abstammung Ur- and Frügeschichte des Haushundes.' *Prähistorische Forschungen.* Vienna.

HAWKES, J. & C. (1944). *Prehistoric Britain.* Chatto & Windus, London.

HECK, L. (1951). 'The Breeding Back of the Aurochs.' *Oryx,* I. London.

HEYN, V. (1888). *The Wanderings of Plants and Animals.*

HITTI, P. R. (1956). *The Arabs: A Short History.* Macmillan, London.

HOWARD, M. N. (1951). 'Dried Cats.' *Man,* 51. London.

HYAMS, E. (1952). *Soil and Civilization.* London.

    (1971). *Plants in the Service of Man.* Dent, London.

KARSTON, R. (1949). 'A totalitarian state of the past: the civilization of the Inca Empire in ancient Peru.' *Proceedings of the Finnish Scientific Society,* XVI. Helsinki.

KERNEY, M. P. (1965). 'Snails and Man in Britain.' *Journal of Conchology*, 26.

KELLER, O. (1909). *Die Antike Tierwelt*, vol. i. Leipzig. (1913). *Die Antike Tierwelt*, vol. ii. Leipzig.

KENCHINGTON, F. E. (1949). *The Commoners' New Forest*. Hutchinson, London.

LANE, Y. (1946). *African Folk Tales*. Lunn, London.

LAUFER, B. (1919). *Sino-Iranica*. Field Columbian Museum, Anthropological Series.

LHOTE, H. (1953). 'Le Cheval et le Chameau dans les peintures et gravures rupestres du Sahara.' *Bulletin de l'Institut français de l'Afrique noir*.

LOW, I. (1967). 'Prospects for sea fish-farming.' *New Scientist*, 6th July 1967. London.

MAJOR, R. H. (1872). *'The Canarian' by Pierre Boutier and Jean le Verrier (1402)*. Hakluyt Society, London.

MALLOWAN, M. E. L. (1956). *Twenty-five Years of Mesopotamian Discovery, 1932–56*. British School of Archaeology in Iraq.

MASSOULARD, E. (1949). *Préhistoire et Protohistoire de L'Egypte*.

MILLS, C. (1820). *History of the Crusades*.

MOMMSEN, T. (1853–6). *The History of Rome*. Everyman edition.

MONTET, P. (1968). 'L'Egypte préhistorique.' In Varagnac (1968).

NACHTSHEIM, H. (1949). *Vom Wildtier zum Haustier*. Berlin.

NASH, C. E. (1968). 'Power stations at sea farms.' *New Scientist*, 14th November 1968. London.

PALLIOT, P. (1664). *La Vraye et Parfaite Science des Armoires*.

PAUL THE DEACON (c. 790). *Historia gentis Langobardorum*.

PENNANT, T. (1776). *British Zoology*.

POLHAUSEN, H. (1949). *Der Gebrauch der Rentiere zum Reiten. Zugleich ein Beitrag zur Frage nach dem Ursprung der Herdentierzucht*. MS.

(1953). *Nachweisbare Ansätze zum Wanderhirtentum in die niederdeutschen Mittelsteinzeit*. MS.

PRESCOTT, W. H. (1843). *The Conquest of Mexico.*

REED, C. A. (1960). *A Review of the Archaeological Evidence on Animal Domestication in the Prehistoric Near East.* Oriental Institute of the University of Chicago, Studies in Ancient Oriental Civilization, No. 31.

REED, C. A. & BRAIDWOOD, R. J. (1960). *Towards the Reconstruction of the environmental sequence in North-west Iraq.*

RÜTIMEYER, L. (1862). *Die Fauna der Phalbauten Sweiz.*

RYDER, M. L. (1969). 'Changes in the fleece of sheep following domestication.' In Ucko & Dimbleby (1969).

SALONEN, A. (1956). 'Hippologica Accadica.' *Annales Academiae Scientiarum Fennicae.* Helsinki.

SIRELIUS, U. T. (1916). 'Über die Art und Zeit der Zähmung des Rentiers.' *Journ. Soc. Finno-Ugrian*, 33 (2).

SPENCE, L. (1945). *Magic Arts in Celtic Britain.* Rider, London.

STONER, C. R. (1950). 'The Feasts of Merit among the Northern Sangtam Tribe of Assam.' *Anthropos*, vol. 45.

(1953). 'The Mithan of Assam.' *Geographical Magazine*, vol. 26. London.

(1957). 'Notes on Religion and Ritual among the Delta Tribes of the Assam Himalayas.' *Anthropos*, vol. 52.

THOMPSON, H. V. (1968). 'Mink, the new British beast of prey.' *The Field*, 5th September 1968. London.

TORQUEMADA, J. DE (1773). *La Monarquia indiana.*

UCKO, P. J., & DIMBLEBY, G. W. (ed.) (1969). *The Domestication and Exploitation of Plants and Animals.* Duckworth, London.

VAN VECHTEN, G. (1921). *The Tiger in the House.*

VARAGNAC, A. (ed.) (1968). *L'Homme avant l'Ecriture.* Paris.

VAUFREY, R. (1951). *Le Palaeolithique et la Mésolithique du Desert de Judée.* Mémoires de l'Institut de la Palaeontologie humaine, No. 24. Paris.

VULSON, M., seigneur de la Colombière (1644). 'Comme le lion est un animal solitaire, aussi le chat est une bête lunatique . . .', in *Livre de la Science héroïque.*

WHITE, G. (1789). *The Natural History of Selborne.*

WHITE FISH AUTHORITY REPORT (1970). *Experiments in Marine Fish Cultivation.*

WOOLLEY, L. (1934). *Ur Excavations II. The Royal Cemetery.* British Museum, London.

ZEUNER, F. E. (1963). *A History of Domesticated Animals.* Hutchinson, London.

Index